Steamboats on the Colorado River, 1852-1916

Steamboats on the Colorado River

1852  1916

Richard E. Lingenfelter

THE UNIVERSITY OF ARIZONA PRESS   TUCSON, ARIZONA

About the Author . . .

Richard E. Lingenfelter, a historian by avocation, has been a professor in residence of geophysics and space physics and astronomy at the University of California, Los Angeles, since 1969. He has written and edited several books on western American history, including *First Through the Grand Canyon, The Newspapers of Nevada, 1858–1958: A History and Bibliography, The Songs of the Gold Rush, The Songs of the American West,* and in 1974 *The Hardrock Miners, A History of the Mining Labor Movement in the American West, 1863–1893.*

THE UNIVERSITY OF ARIZONA PRESS

Copyright © 1978

The Arizona Board of Regents
All Rights Reserved
Manufactured in the U.S.A.

Library of Congress Cataloging in Publication Data

Lingenfelter, Richard E
   Steamboats on the Colorado River, 1852–1916.

   Bibliography: p.
   Includes index.
   1. Steam-navigation—Colorado River—History.
2. Colorado River—navigation—History. 3. River steamers—Colorado River—History. I. Title.
VM623.L56   386'.3'097913   78-16241
ISBN 0-8165-0650-7
ISBN 0-8165-0567-5 pbk.

For permission to use the illustrations contained in this volume we wish to credit the Arizona Department of Library, Archives and Public Records, p. 26; the Arizona Historical Society Library, pp. 25, 28, 39, 87, 89, 92–94; The Bancroft Library, pp. 32, 54, 57, 59, 70, 79, 178; Barbara Baldwin Ekker, p. 119; the Church Archives Historical Department, The Church of Jesus Christ of Latter-day Saints, p. 48; Mrs. Edwin Wilcox, pp. 107, 116; the Engineering Societies Library, p. 77; H. E. Huntington Library, San Marino, California, pp. 15, 45, 46, 75, 83, 90, 170, 186; Historical Collection, Title Insurance & Trust Co., San Diego, California, pp. 55, 56, 63, 140; the Map Library, University of California, Los Angeles, p. 61; the Nevada Historical Society, Reno, pp. 50, 85; the New York Public Library, pp. 124–26; Otis Marston, pp. 104, 121; Sharlot Hall Historical Society, pp. 30, 52, 67; Southern Pacific Railroad Company, pp. 72, 147; Southwest Museum, pp. 38, 42, 134; Special Collections Library, the University of Arizona, p. 160; Special Collections Library, University of California, Los Angeles, pp. xvi, 3, 4, 6–8, 13, 14, 18, 20, 22, 27, 43, 58, 65, 101–2, 111, 137, 141, 143–44, 146, 148, 150–55, 180; Stimson Photo Collection, Wyoming State Archives and Historical Department, p. 120; Tex McClatchy, p. 133; the Title Insurance Company, Los Angeles, p. 123; the University of California Library, Berkeley, p. 98; the U.S. Geological Survey, p. 132; the U.S. Military Academy, p. 17; the U.S. National Archives and Records Services, pp. 69, 96, 114, 128, 130, 157–58; and Wilbur Rusho, p. 131.

Cover: *Lithograph of the steamboat* Explorer, *by courtesy of the Special Collections Library, the University of California, Los Angeles.*

Title Page: *Lithograph of the steamer landing at Fort Yuma, by courtesy of the Arizona Historical Society Library.*

For Kendale

1962 – 1978

Contents

List of Illustrations	ix	Appendixes	
Preface	xiii	A. Steamboats	161
Opening the River	1	B. Operators	164
The Arizona Fleet	31	C. Table of Distances	167
Through Progress of the Railroads	73	Notes to the Chapters	171
Steamboats in the Canyons	105	Bibliography	181
Closing the River	135	Index	187

Illustrations

River Steamers

Steamboat *Explorer* in Cane Break Canyon	xvi
Johnson's First Steamer, the Side-Wheeler *General Jesup*	13
The *Explorer* Steaming Past Chimney Peak	18
The *Explorer* Dwarfed by Mohave Canyon	22
The *Gila, Cocopah,* and *Barge No. 3* Lined Up at Yuma	30
The *Colorado* (II) Opposite Chimney Peak	42
The *Mohave* (I) Built in 1864	46
The *Cocopah* (II) Built in 1867	52
The *Colorado* (II) in the Dry Dock at Port Isabel	54
The *Gila,* the Most Durable Boat on the Colorado	55
Largest Colorado Steamboat, the *Mohave* (II)	56
Ads for the Steamers *Newbern* and *Montana*	57, 59
Barge No. 2 Alongside the *Mohave* (I)	63
The *Gila* at Castle Dome Landing	67
The *Gila* With a Bargeload of Coal	69
Ad for the "Palatial" Riverboat *Mohave* (II)	79
The *Gila* Beneath the Santa Fe's Cantilever Bridge	83
The Rival Steamboat *St. Vallier*	92
The *Cochan* Rebuilt From the *Gila*	93
The *Searchlight,* Another Rival Steamboat	96
Tied Up at Yuma, the *St. Vallier, Cochan,* and *Silas J. Lewis*	102
Lute H. Johnson on the Little *Major Powell*	104
Canyon Steamer *Undine*	111
The *Cliff Dweller* at Green River Bridge, Utah	116
Steamer *Comet* With Construction Crew	121
The *Charles H. Spencer* Under Construction in Glen Canyon	130
The *Charles H. Spencer* With Crew	131
The *Charles H. Spencer* Left to Rot	132
Silhouette of the *Cochan* at Twilight	134
The *St. Vallier* Attempting to Close the Intake	148

x *Illustrations*

The *Searchlight* and Her Barge Stranded in the Dry Channel	153
The *Gila* With Two Barges Loading at Yuma	160
The *St. Vallier* Awaiting Cargo	170
The *Gila* Above Southern Pacific Railroad Bridge	178
The *Searchlight* at Imperial Canal Intake	180
Last Survivor of the Fleet, the *Searchlight*	186

Other Boats

Barge No. 3 With the *Gila* and *Cocopah*	30
Barge No. 4 Being Towed Through Red Rock Gate	45
The Ill-fated *Montana*	58
Barge No. 2 With the *Mohave* (I)	63
The *Aztec* Passing Through Railroad Swing Bridge	87
The *Aztec* Packed With Excursionists	89
Unidentified Boat Used as a Ferry at Needles	90
Ad for the "Steamers" *Iola* and *Hercules*	98
Plan of the *Advance* Gold Dredge	101
The *Silas J. Lewis* With the *St. Vallier* and *Cochan*	102
The Ungainly *City of Moab*	114
Moab Garage's "Big Boat"	119
The *Hoskaninni* Being Built at Camp Stone	125
The *Hoskaninni* at Work	126
Tex McClatchy's *Canyon King*	133
Ives's Dredge, the *Alpha*	137
Suction Dredge *Beta*	143
The *Beta* and *Alpha* Cutting New Intakes	144

Design for Dredge *Delta*	151
The *Delta* Squeezing Through Swingspan Bridge	152
The *Advance* Working on Laguna Dam	157

Personalities

Yuma Indians	3
George A. Johnson	8
David C. Robinson	15
Lt. Joseph Christmas Ives	17
Mohave Chiefs Cairook and Iretaba	27
Isaac Polhamus, Jr.	28
Johnny Moss With Piute Chief Tercherrum	32
William Harrison Hardy	39
Thomas E. Trueworthy	43
Anson Call	48
John Alexander Mellon	65
Isaac Polhamus, Jr.	75
Robert B. Stanton	124

Places and Events

Fort Yuma and Jaeger's Ferry in 1852	6
Robinson's Landing at the Mouth of the Colorado	14
A Mohave Rancheria	20
Steamboat Landing and Cable Ferry at Fort Yuma	25
Fort Mohave	26

Ehrenberg, Main Upriver Port	38
The Ruins of Callville	50
Imaginary Port of Piute City	70
Locomotive Crossing the Colorado at Yuma	72
Stamp Mill at Eldorado Canyon	77
Boom Camp of Searchlight, Nevada	85
Passengers Aboard a Steamboat	94
Riverport of Green River, Utah	107
Steamer Landing at Green River, Wyoming	120
Gold Miners in Glen Canyon	123
Spencer's Contraption for Extracting Gold From Shale	128
Imperial Town Lot Auction	140
Fresno Scrapers Digging Connecting Canals	141
Mexicali, Flooded in 1905	146
Flooding of Railroad Tracks in Salton Sink	147
Carloads of Rock Used in Damming the Intake	150
Final Closing of the Break in 1907	154
Land Scarred by the Man-Made Floods	155
Completion of Laguna Dam in 1909	158

Maps

Whipple's Map of Colorado River Crossings, 1849	4
Hardy's Erroneous Map of the Colorado Estuary	7
Steamboat Landings on the Colorado River, 1850s	10
Steamboat Landings on the Colorado River, 1860s and 1870s	34
Chart of the Mouth of the Colorado, 1873–1875	61
Steamboat Landings on the Colorado River, 1880s–1900s	80
Steamboat Landings on the Upper Colorado and Green Rivers, 1890s–1910s	108
Colorado River During the Flooding of Imperial Valley, 1905–1907	138

Preface

For nearly fifty years after the 1849 gold rush, paddle-wheel steamboats provided the cheapest and most efficient form of transportation in the West. During this time the Columbia, the Sacramento-San Joaquin and the Colorado rivers became the main thoroughfares for opening the interior to settlement and development. Within a few years of one another each of these rivers was supporting a thriving steamboat business. Their subsequent courses ran roughly parallel. From a brisk period of innovation, competition and expansion, they passed through a stifling, but profitable, time of monopolization to a final period of slow decline and obsolescence with the coming of the railroads and the gasoline engine. Despite the relative isolation of the great rivers, especially the Colorado which was more than two thousand miles by water from the Sacramento, there was interchange of men and boats between the rivers as well. Steamboats and men most commonly moved from the Columbia to the Sacramento or from the Sacramento to the Colorado; however, one of the pioneers of the Colorado River business left that river to head the monopoly on the Sacramento, and a gang of Columbia River ship's carpenters built what was to become the last surviving steamboat on the Colorado.

Thus the course of steam navigation on the rivers was related to and occasionally intertwined with the overall development of transportation in the West, while at the same time each was in its own way unique and separate. The history of western steamboating has been sketched by Oscar Winther in *The Transportation Frontier* and Harry Drago in *The Steamboaters*. The history of Columbia River steamboating has been thoroughly treated by Randall V. Mills in *Stern-Wheelers Up Columbia: A Century of Steamboating in the Oregon Country* and the Sacramento-San Joaquin by Jerry MacMullen in *Paddle-Wheel Days in California*.

The following pages will attempt to tell the story of steam navigation on the Colorado and its tributaries, reaching from the

xiv *Preface*

Gulf of California to the Green River in Wyoming, and from the launching of the first steamer on its waters in 1852 to the loss of the last in 1916. Only the early part of this story has previously been told, principally by Francis H. Leavitt in his article "Steam Navigation on the Colorado," published in the *California Historical Society Quarterly* in 1943, by Arthur Woodward in his 1955 book, *Feud on the Colorado,* and by Richard N. Coolidge in his 1963 thesis at San Diego State College, entitled "History of the Colorado River During the Steamboat Era."

Gathering the material for this history would not have been possible without the generous aid of James Mink, Brooke Whiting and the staff of Special Collections, the University of California, Los Angeles, and Edith Fuller and the staff of Interlibrary Loans of the Research Library of the same university; John Barr Tompkins of the Bancroft Library, University of California, Berkeley; Margaret Sparks Bret Harte of the Arizona Historical Society Library, Tucson; Richard A. Ploch and Jack Latham of the Special Collections Library, University of Arizona, Tucson; Wilma Smallwood of the Arizona State Department of Library and Archives, Phoenix; Doris Heap and Ionne Ladd of the Sharlot Hall Museum, Prescott, Arizona; Marylou Wilkey and Cliff Traszer of the Yuma County Historical Society, Yuma, Arizona; Frances E. Thomas of the Yuma City-County Library; John Murphy and Kermit Edmonds of the Mohave Pioneers' Historical Society, Kingman, Arizona; William H. Haught of the Yuma Territorial Prison State Historic Park; Lee R. Burtis of the California Historical Society Library, San Francisco; Miriam T. Pike of the California Section, California State Library, Sacramento; Ruth M. Christensen of the Southwest Museum Library, Los Angeles; Virginia Renner of the H. E. Huntington Library and Art Gallery, San Marino, California; Wayne Fabert of the San Diego Historical Society, San Diego; Hank R. Wilde of the Needles Museum Association, Needles, California; Jeannette Hargrave, Librarian of the U.S. Army Corps of Engineers, Los Angeles District; Dale Reed of the State Historical Society of Colorado, Denver; Hazel Lundberg of the Western History Section, Denver Public Library; Eslie Cann of the Nevada Historical Society, Reno; Pamela Crowell of the Nevada State Museum, Carson City; Celesta Lowe of Special Collections, University of Nevada Library, Las Vegas; Robert Armstrong of Special Collections, University of Nevada Library, Reno; Paul R. Rugen and Jean R. McNiece of the Manuscript Division, New York Public Library; Melvin T. Smith of the Utah State Historical Society, Salt Lake City; Eleen T. Williams of the Daggett County Historical Society, Dutch John, Utah; Donald Mosholder and Michael Goldman of the U.S. National Archives and Records Service, Washington, D.C.; Charles G. Roundy of the Western History Research Center, University of Wyoming Library, Laramie; and Henry F. Chadey of the Sweetwater County Museum, Green River, Wyoming.

I am also particularly indebted to Pearl Baker of Green River, Utah; H. Elwyn Blake of Albuquerque; Juanita Brooks of Salt Lake City; Dennis G. Casebier of Norco, California; Carl A. Gaensslen of Green River, Wyoming; Otis Marston of Berkeley; David Myrick of San Francisco; Adrian Reynolds of Green

River, Wyoming; Dwight L. Smith of Miami University, Oxford, Ohio; Samuel J. Taylor of Moab, Utah; and Roscoe Willson of Phoenix.

For the use of early photographs I am grateful to many of the above as well as to Barbara Ekker of Hanksville, Utah; Wilbur L. Rusho of Salt Lake City; Mrs. Edwin Wilcox of Green River, Utah; Tex McClatchy of Moab, Utah; Office of the Church Historian, Salt Lake City; Engineering Societies Library, New York; Title Insurance and Trust Company, San Diego; Title Insurance and Trust Company, Los Angeles; U.S. Geological Survey, Denver; U.S. Military Academy, West Point; U.S. Reclamation Bureau, Boulder City, Nevada and Wyoming State Archives and Historical Department, Cheyenne.

Finally, thanks go to the University of Arizona Press for effecting publication of this volume.

R. E. L.

The steamboat *Explorer* venturing through Cane Break Canyon in 1858, sketched by F. W. Egloffstein, topographer for the Ives expedition mapping the Colorado River.

Opening the River

In November 1852 a homely little steam tug, the *Uncle Sam,* was launched on the muddy waters at the mouth of the Colorado River. As a handful of Cocopahs, Sonorans and Yankees watched with amusement, sparks popped from her firebox, mesquite smoke belched from her stack, and her hand-me-down engine shuddered and clanked. Finally, with a straining creak her paddles started to stir the cloudy water, and, hesitantly, she pulled away from the bank to head up the unknown river.

From such modest beginnings the era of steam navigation on the Colorado River began—an era that would bring the opening of mines and settlements all along the river. Eventually a swift fleet of stern-wheelers would run more than six hundred miles up the river from the Gulf of California to the rapids just below the Grand Canyon. Even in the canyons above, a few sturdy little craft would brave every patch of smooth water all the way to Wyoming, sixteen hundred miles from the gulf. For more than half a century the valleys and canyons of the Colorado would echo to the shrill of the steam whistle, the throb of the pistons and the song of the leadsman.

The river steamers were to become the very lifeline of Arizona, carrying in soldiers, miners, ranchers and merchants, and all of their rations, tools, furniture and wares; and carrying out the wealth of the mines—millions in gold, silver, copper and lead. The story of steamboating on the Colorado River is an important part of the story of the opening of Arizona.[1]

The Colorado was a bountiful home for the Cocopah, Yuma, Mohave and others long before Francisco de Ulloa first sighted its muddy estuary in 1539, and the subsequent sporadic intrusions of Spanish explorers, Franciscan missionaries and Yankee fur trappers had little impact on river life. But suddenly the great gold rush to California in 1849 brought a massive invasion of the river country and with it the subjugation of the river tribes.

It was as a part of this invasion that steam navigation of the Colorado River began. Late in 1849 an enterprising Tennessean, a Dr. Lincoln, settled on the Colorado just below the junction of the Gila, establishing a ferry at the Yuma Crossing to exact a toll from every westering argonaut headed for the gold fields by the southern route. His business was so good that a gang of Texas scalp hunters, led by John Glanton, soon muscled in for a share. It came to a tragic end the following spring, however, when the Yuma Indians also tried to get a little of the business with a rival ferry. Glanton's gang broke up their boat and drowned one of their men. The Yumas promptly retaliated, demolishing Glanton's ferry and killing all but three of the gang.

When the survivors reached San Diego, their tales of the massacre excited a clamor among pot-valiant patriots for a punitive expedition against the Indians. The governor of California appointed state Q.M. Gen. J. C. Morehead to lead the foray. With a comical force of several dozen idlers and drunks rounded up by the county sheriff, Morehead set out across the desert for the Yuma Crossing in the heat of the summer. Others, in the meantime, began petitioning the federal government to establish a permanent military post at the crossing to protect overland travelers from similar attacks.[2]

At the same time exaggerated accounts of the profits of the ferry business, circulating with the news of the massacre, lured a young San Francisco stevedore, George Alonzo Johnson, to the river. Johnson, a twenty-four-year-old up-state New Yorker, who had worked on the Great Lakes steamers before coming to the gold fields, was destined to play a leading role in steamboating on the Colorado. He and several partners, including Benjamin M. Hartshorne, who later joined him in the steamboat business, sailed from San Francisco in June 1850 to reestablish the ferry. They reached the river ahead of Morehead and built a stockade and two ferryboats under the wary eye of the Yumas.[3]

Soon after, Morehead's expedition came straggling in off the desert. The general, angered at having been beaten out of the ferry business, tried to take out his frustration on the Yumas, but his men soon turned their attention more profitably to Sonorans returning from the gold field, disarming them and relieving them of their gold dust. Morehead finally headed back to the coast just before three companies of the 2nd U.S. Infantry arrived to establish a permanent post.[4]

The troops, commanded by Maj. Samuel P. Heintzelman, reached the river on 1 December 1850, setting up Camp Independence about six miles below the ferry. The Yumas offered only token resistance to the invading troops before agreeing to a peace. The major was a cagy Pennsylvania German, and, as soon as military affairs were settled, he too got an "itching" to share in the ferry profits. Thus in March 1851 he established a new post, Fort Yuma, on a low mesa overlooking the ferry crossing, and he laid out a military reservation which completely encompassed the ferry property. When Johnson protested, the major simply suggested that he sell out. As the ferry had not proved as profitable as Johnson had hoped, he and most of his partners did just that, selling their shares for about $3,000 and a mule each. This left

Yuma Indians living at the emigrant crossing profited by trade with the argonauts. Lithograph by Balduin Möllhausen.

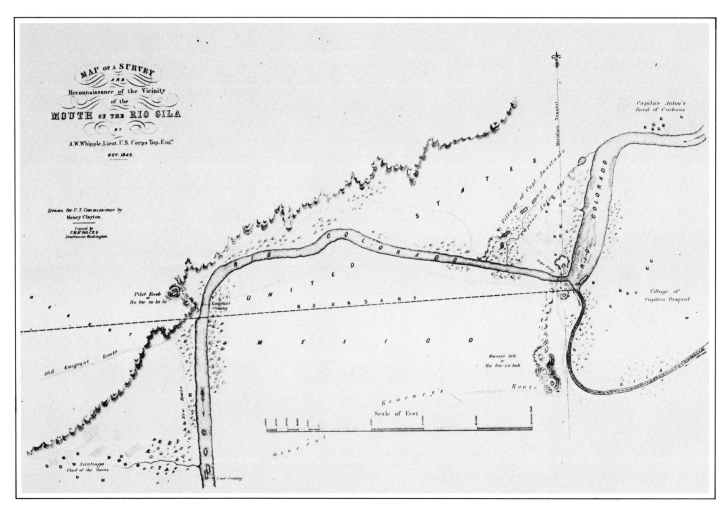

Whipple's map of the Colorado River crossings in 1849 shows Lincoln's rope ferry just below the mouth of the Gila and the rival crossings downriver.

only Louis J. F. Jaeger, William J. Ankrim and the major, owners of the ferry. Johnson returned to San Francisco, but within a year he was drawn back to the river.[5]

Fort Yuma quickly became known as the hell hole of the West, for its blistering hot summers, its desolate loneliness, its abominable quarters and its starvation rations. Jokes seemed to offer the only respite against the heat, and the soldiers accosted newcomers with the tale of a comrade who died and went to hell but came back the next day for his blankets. Jacals, made of upright mesquite sticks dabbed with mud and a thatching of arrowwood, provided the soldiers' only shelter from the wind, dust and sun. The officers did little better, stretching their tents among the few crumbling walls of the old mission La Purísima Concepción, which stood on the mesa, a lonely reminder of an ill-fated Spanish attempt to settle the river in 1780. More livable adobe quarters were finally begun in 1855.[6]

The greatest difficulty facing the garrison in those early years was the chronically short supply of rations. The struggle to alleviate this problem led to commencement of steam navigation on the Colorado, but it was a slow and faltering process fraught with frustration. The first attempt to supply the fort was by wagon and pack mule across the mountains and desert from San Diego, a distance of more than two hundred miles. Freighting costs of $500 a ton, however, made this route much too expensive, so the quartermaster looked to the sea as an alternative. The latest map of the lower Colorado was one surveyed by British naval Lt. R. W. H. Hardy in 1826 while searching for pearling grounds in the Gulf of California. Hardy placed the junction of the Gila, where the fort was located, only twenty-five miles above the mouth of the Colorado within easy reach of small seagoing craft. Thus the quartermaster dispatched Lt. George Horatio Derby, the witty "Squibob" of later literary fame, to the mouth of the river with a cargo of 10,000 rations in the little schooner *Invincible*, commanded by Captain Alfred H. Wilcox.[7]

Derby sailed into the mouth of the river on Christmas Eve of 1850 and soon worked the vessel up to the point Hardy had shown as the junction of the Gila. But instead of finding the fort, he discovered that the lieutenant had mistaken a slough for the main channel of the Colorado and the Colorado itself for the Gila. Captain Wilcox refused to take the *Invincible* any farther up the river in search of the fort, so after firing guns in the hope the soldiers might hear them, they sent Cocopahs to carry word to the fort, which proved to be 120 miles farther upriver. On learning of their arrival Major Heintzelman sent wagons down the Sonora side to bring up the much-needed supplies.[8]

In the meantime the *Invincible* was being badly buffeted by the shifting currents and the daily tidal bore—a six-foot wall of water that came in with a roar that could be heard for miles and a force that could capsize or ground a small vessel. Only at the mouth of the Ganges River in India was there a tidal bore said to be as treacherous as this. After going aground a couple of times and losing both anchors, Wilcox decided that he could not risk waiting for the wagons. As a result they left the rations piled on the Sonora bank and set sail.[9]

Fort Yuma and Jaeger's Ferry, sketched by John Russell Bartlett, who passed through in June 1852 while preparations were being made to try supplying the fort by steamer.

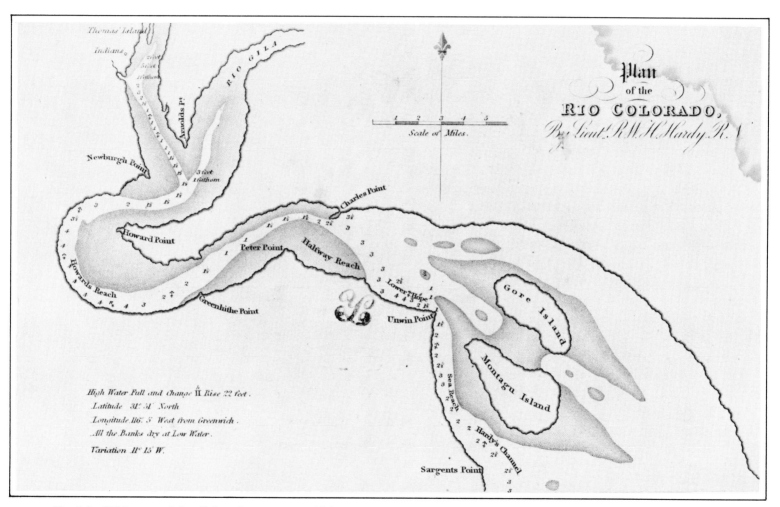

Hardy's 1826 map of the Colorado estuary, which erroneously showed the mouth of the Gila only a short distance up-river and led to the belief that Fort Yuma could be directly supplied by seagoing vessels. (Depths are shown in fathoms.)

This arrangement was even worse than hauling the supplies overland from San Diego. Moreover, it was a "palpable violation" of the Treaty of Guadalupe Hildago for U.S. troops to haul off goods landed in Mexico. During his brief stay, however, Derby clearly saw that the solution to the problem was to put a steamboat on the river. He specifically recommended "a small sternwheel steamer with a powerful engine and thick bottom . . . eighteen or twenty feet beam, drawing two and a half to three feet of water." These specifications were, in fact, a perfect description of the boats ultimately found most suitable for the river, but Derby's recommendations were ignored and his conclusions were reached again only after years of unsuccessful experimentation. Such a boat, Derby claimed, could "carry more to the post in twenty-four hours than a hundred wagons could transport in a week." Heintzelman also endorsed the suggestion but no action was taken.[10]

Early in June 1851 the shortage of rations became so critical that Heintzelman withdrew the bulk of his troops, leaving only eleven men to hold the post until new stores could be brought in. After six months without relief, even they ran out of grub and were forced to abandon the fort altogether. Relief, however, was finally on the way.[11]

In September George Johnson and Ben Hartshorne took a contract to supply the fort. Ignoring Derby's recommendations, they proposed to lighter the supplies up the river on flatboats with poles. They arrived at the estuary in February 1852 in the U.S. schooner *Sierra Nevada* under command of Captain Wilcox.

George A. Johnson came to the Colorado in 1850 to run the ferry and a few years later started a fleet of river steamers that monopolized the river trade for decades.

In anticipation of their arrival Heintzelman reoccupied Fort Yuma that same month and sent a detachment of troops down the river to meet them. Johnson quickly assembled his two flatboats—each 18 by 50 feet with 3-foot draft capable of holding 30 tons. As before, this attempt was plagued with difficulty. The soldiers were attacked and turned back by hostile Yumas, and no sooner was the first flatboat loaded than it swamped and sank, cargo and all a total loss. Johnson and his crew labored valiantly, working the remaining flatboat up the river against the perverse currents of the Colorado, but the troops at the fort consumed the rations faster than he could deliver them. Once again the wagons were sent down to assist. Even then it took four months to get the last of the supplies up the river.[12]

When all else had clearly failed Derby's call for a steamboat was finally heeded. In June 1852 the quartermaster let a new supply contract for the fort—this time to Captain James Turnbull, "an energetic, smooth talking, little fellow" from Benicia, California. Turnbull purchased a small steam tug, had it broken down into sections and shipped with his first cargo of supplies from San Francisco on the schooner *Capacity,* under Captain Driscoll. Anchoring the schooner near the head of tidewater early in September, Turnbull began reassembling the tug. The work went slower than expected, taking over two months, while the garrison waited anxiously for much-needed provisions.[13]

In mid-November 1852 the first Colorado steamboat was finally launched on the muddy waters. Turnbull proudly christened her the *Uncle Sam,* but her appearance reflected poorly on the name. Her little double-pointed hull was only 65 feet long with a 16-foot beam, and 3.5 feet deep—scarcely larger than Johnson's flatboat; her deck, devoid of cabin or wheelhouse, sported only a makeshift 20-horsepower locomotive engine, modestly draped in canvas to shield it from the silty spray of her two side paddles as they churned against the swift currents.[14]

With a crew of two, Captain Turnbull and engineer Phillips, and a few curious passengers, Captain Driscoll, two Cocopah chiefs and a Yuma, the little steamer started bravely up the river on 18 November, her deck heaped with some thirty-five tons of freight. Her progress was frustratingly slow as Turnbull guided her cautiously through the uncertain channels and shifting bars. Even under a full head of steam her engine was barely a match for the river and several hours each day were spent tied up at the bank while all hands foraged for wood to fire her boiler. Trouble with the boiler further delayed them two and a half days. As if all this were not trial enough, they were also frightfully shaken up by an earthquake. Though the *Uncle Sam* came through relatively unaffected, the quake left the *Capacity* high and dry at the estuary and at the fort it sent the frightened troops scampering onto the parade ground to the great amusement of the quake-weary Yumas.[15]

The soldiers, now much more concerned with aftershocks of the quake, had nearly given up looking for the steamboat when she finally nosed into the landing at Fort Yuma early in December. A crowd of soldiers and Indians, however, quickly gathered around her. "She is almost as great a curiosity," one noted, "as

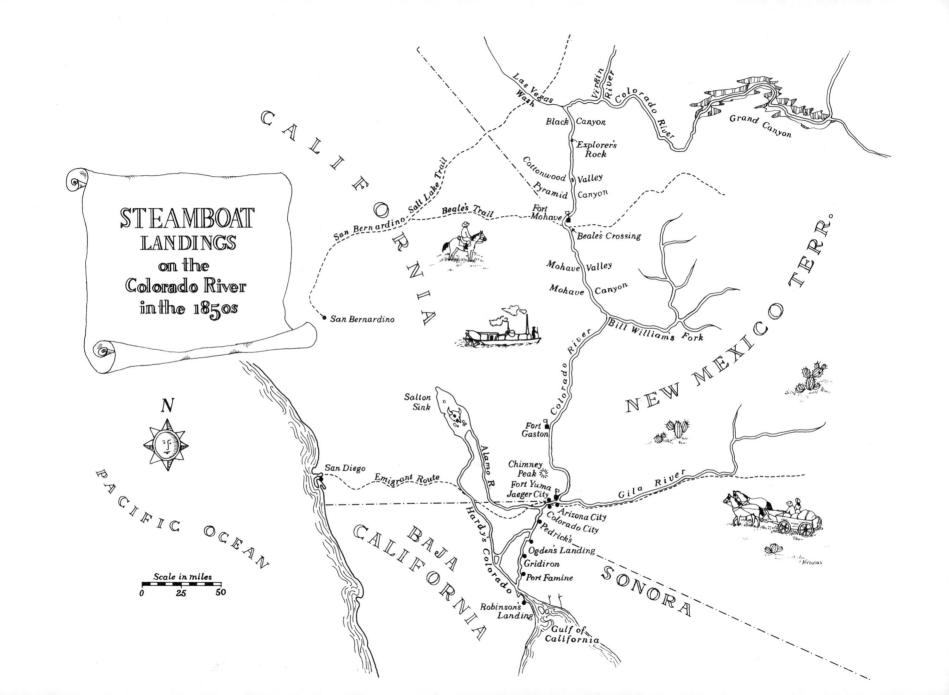

the steamer Fulton had on the Hudson River." Fledgling that she was, she had achieved her goal and the press on the coast soon spread the word that "the practicability of navigating the Colorado by steam is established beyond doubt." Indeed a new era had opened on the river, but the problem of supplying the fort was not yet solved.[16]

Before heading back down for another load of supplies, Captain Turnbull treated the off-duty officers to a short excursion on up the Colorado and Gila above the fort. As one of the excursionists later remarked, "The trip was rather pleasant than otherwise, more on account of its novelty than anything else, I surmise, for we got pretty well sprinkled during the voyage." Had it been summer rather than winter this might have been appreciated.[17]

On her first trip upriver the *Uncle Sam* had taken fifteen days to make the 120 miles from the schooner to the fort, but eventually she was making a round trip in twelve days. Even at this rate, however, the *Capacity* was not unloaded until mid-April —having been moored on the river for eight months. By January it had already become obvious to Turnbull that the poor little steamer was sadly underpowered. Leaving Captain Driscoll to finish the unloading, Turnbull headed overland to San Francisco to get a more powerful engine. He returned to the river in mid-May aboard the schooner *General Patterson* with a new engine and another cargo of provisions for the fort. But sad news awaited him.[18]

After unloading the *Capacity,* the *Uncle Sam* had been tied up at the old ferry crossing six miles below the fort to be overhauled and made ready for the new machinery. There, shortly before Turnbull's return, someone forgot to put a bilge plug in tight and she filled with water and sank. Several men from the fort went down to help raise her. They worked tirelessly for two days before she broke her moorings and disappeared in the swift muddy current. The first steamboat on the Colorado thus went to an early grave.[19]

Despite the loss Turnbull, vowing he had "not yet given up," headed back to San Francisco once again—this time to get a new hull. Once again the army had to send the wagons down to the rescue. By the time Turnbull had reached San Francisco, however, he had had second thoughts about throwing any more money into the Colorado steamboat business. Thus he quickly and quietly vanished from the scene, leaving his creditors no recourse but to fire broadsides at him through the press, advertising him as "a mean, contracted hypocrite, not worthy of any gentleman's confidence." Turnbull was his own best advertisement, however, and he soon became "wellknown around Mazatlan where he ran a little stern-wheel boat for years, and better known for his attempt to build canals, etc. down there on pure jawbone—without any money at all!"[20]

His faults aside, Turnbull had shown that a river steamer offered the only practical solution to the problem of supplying Fort Yuma. Even George Johnson was convinced. That fall Johnson sailed for the river on the brig *General Viel* with a new supply contract in his pocket, and in the hold, the sections of a new steamer he had purchased in partnership with Ben Hartshorne and Captain Alfred H. Wilcox. This new boat was a side-wheeler like the *Uncle Sam,* but she was much larger and more powerful than her predecessor. She had a 50-horsepower engine, measured 104 feet from stem to stern, and 17 feet at the beam—27 feet including her paddle guards—and carried 50 tons on only 30 inches of water. Johnson thoughtfully named her the *General Jesup* in honor of the U.S. quartermaster general whose business made her possible. She was reassembled in the estuary and reached Fort Yuma with her first cargo on 18 January 1854.[21]

Unlike the *Uncle Sam,* the *General Jesup* was an immediate financial success, clearly demonstrating the economic feasibility of steam navigation on the Colorado. She made round trips from the estuary to the fort in only four to five days with 50-ton cargoes paying $75 a ton, to gross nearly $4,000 a trip, or $20,000 a month, during the busiest season. Johnson, in fact, pushed her for all she was worth. In August 1854, straining to make time against the current, her boiler exploded, killing the engineer Jackson and seriously scalding two others. New machinery was sent down from San Francisco, and she was running again by December. Johnson's new engineer, David Neahr, kept a much closer eye on the boiler and lived to a ripe old age.[22]

The faster times made by the *General Jesup* were due not only to her more powerful engine but also to the establishment of woodyards which eliminated the time-consuming quests for firewood. Aside from scattered Cocopah rancherias, the woodyards were the only settlements on the river below the ferry. They were roughly spaced about a day's voyage—some thirty miles—apart so the steamer could be loaded with wood while she was tied up for the night. The ever-changing channel of the Colorado made navigation at night too hazardous. The first yard above the mouth of the river was ominously known as Port Famine; above that was the Gridiron, then Ogden's Landing, run by one of Johnson's old ferry partners, and finally old John Pedrick's, just above the boundary line—the first landing in the United States. Though most of the yards were on Mexican soil, the owners were all Yankees. They hired Cocopahs to cut and haul the wood to the river and were said to make as much as $5,000 a year from the business.[23]

At the mouth of the river there was one other settlement—if it could be called such—known as Robinson's Landing. It was on the mud flats on the Baja California side. Here atop stilts, since it was awash at high tide, stood a solitary shed, pretentiously dubbed the Colorado Hotel. Though gulls and pelicans were its only regular customers, it became an indispensable landmark for seagoing vessels rendezvousing with the steamer. It was built by David C. Robinson, who had first come to the river as mate on the *Invincible* in 1850 and returned to the river with Wilcox and Johnson in 1853. "Old Rob," as he was affectionately known, served as mate and later captain on the river steamers. He kept the "hotel" as a base for his intermittant search for a cache of gold, said to have been lost by Count Raousett de Boulbon's ill-fated Sonoran filibustering expedition when their boat sank on Hardy's Colorado. Freight was rarely landed here, however, since Johnson transferred his cargoes directly from seagoing vessels in mid-stream to avoid paying Mexican customs duty.[24]

Johnson's first steamboat, the *General Jesup,* shown in this 1854 lithograph, and Turnbull's earlier *Uncle Sam* were the only side-wheel steamers ever run on the Colorado. Though Johnson worked the *General Jesup* upriver as far as the present site of Davis Dam, he soon concluded that sternwheelers were much better suited to the swift currents and shifting bars of the Colorado.

Robinson's Landing at the mouth of the Colorado was the transfer point for freight and passengers from sea-going vessels to the river steamers in the 1850s and was also the staging point for Captain Robinson's intermittant quests for an elusive treasure said to have been buried by a Sonora filibustering expedition. Lithograph by J. J. Young from a photograph by Lt. J. C. Ives.

David C. Robinson, who came to the Colorado in 1850 with Derby, returned with Johnson to become a captain of the river steamers.

Commerce on the Colorado developed rapidly in the next few years. With adequate shipping assured, miners, ranchers and merchants settled along the river, bringing, in turn, more trade to Johnson's steamer. In 1853 the mess cook at Fort Yuma, Mrs. Bowman, affectionately known as "The Great Western," laid claim to the land opposite the fort and built the first house in what is today Yuma. She sold it the following year to the fort sutler, George F. Hooper, for a store. As business picked up a rival trader and a couple saloonkeepers also set up shop. When a post office was established in 1858, the cluster of adobes became known as Arizona City. Two rival settlements were started about a mile downstream at Jaeger's ferry. The largest, Jaeger City, on the California side, for a time surpassed Arizona City, boasting a hotel, two stores, two blacksmiths, the Overland Mail offices and several other dwellings. Across the river from it Charles Poston surveyed a townsite which he dubbed Colorado City, giving Jaeger an interest just for ferrying his party across the river. Poston sold the townsite for $20,000 in San Francisco and the customshouse was built there. Both rivals were almost completely swept away by flood waters in January 1862. Colorado City was rebuilt, but finally merged into Arizona City, leaving it the unchallenged entrepôt to Arizona.[25]

On the rich bottomlands above the fort a couple ranches were started in 1853 to supply beef and barley to the army. Gold was found along the river late that same year, and the following year copper was discovered some forty miles above. The copper mine gave Johnson his first paying cargo on the downriver trips as the ore was shipped out for smelting, but he and his partners lost this revenue when they all too eagerly bought the mine. Major Heintzelman, always looking for a little profit on the side, also went into mining, reopening several old Mexican mines in the recently acquired Gadsden Purchase. These, too, began shipping out ore and bringing in machinery via the Colorado.[26]

By the summer of 1855 the river trade was so brisk that the *General Jesup* could no longer keep up. Four vessels were lying in the estuary while their captains ranted that they were being unloaded only by "teaspoonfuls." Johnson, however, had already realized that the *Jesup* needed help and in the hold of one of the vessels were the pieces of a second steamer. She was larger and more powerful yet—120 feet long, with an 80-horsepower engine and a capacity of 60 to 70 tons. She was also the first stern-wheeler—as Johnson had finally come to fully appreciate Derby's conclusion that a stern-wheeler was best suited for fighting against the swift currents and maneuvering among the shifting bars of the Colorado. His growing respect for the river also prompted Johnson to name the vessel for her. By December the *Colorado* was in service, steaming up the river "like a streak of lightning," the swiftest boat yet put on the river.[27]

With two boats now on the river, Johnson began looking for ways to expand his business. Talk of sending a steamer on up the river to open trade with the Mormons in Utah had first begun soon after the *Uncle Sam* reached Fort Yuma. Brigham Young was also interested in opening a "sea route" for emigration to Zion by way of the Colorado. Moreover, Antoine Leroux, an old trapper who had rafted down the Colorado from the mouth of the Virgin River in 1837, claimed it was navigable by steamboat all the way. Johnson soon set out to test this claim.[28]

In 1856 Johnson talked the California legislature into passing a joint resolution favoring an expedition up the river, and with this he persuaded the Secretary of War, Jefferson Davis, to ask the Congress for $70,000 to fund it. The appropriation passed, but with the change in administrations after the election that year, the new secretary, John B. Floyd, named one of his inlaws, a young lieutenant in the Corps of Topographical Engineers, Joseph Christmas Ives, to undertake the expedition. Ives, a rather foppish, self-aggrandizing fellow, saw the opportunity as "the event of his life." Johnson, expecting to have the honor himself, was angry over Ives' appointment, and when Ives further refused even to hire his boat he was furious. He had offered either the *Jesup* for $3,500 a month or the *Colorado* for $4,500, a fair rate considering their monthly take hauling freight. But Ives, claiming the rate was exorbitant, convinced his superiors that he could build a boat of his own for less. When he reached the river, however, Ives did accept Johnson's offer of Captain Robinson's services as pilot, and for all his vanity Ives freely acknowledged that to Robinson was "due, in great measure, the successful ascent of the Colorado."[29]

During the summer of 1857 Ives had a small iron-hulled stern-wheeler, the *Explorer,* built in Philadelphia and tested on the placid Delaware before being dismantled for shipment to the Colorado. Arriving at the river at the end of November, Ives set

Lt. Joseph Christmas Ives, a foppish young West Pointer given command of the expedition to find the head of navigation of the Colorado, so angered Captain Johnson by refusing to charter one of his boats that Johnson set out in the *General Jesup* ahead of Ives to claim for himself the honor of first exploring the river.

up camp at Robinson's Landing. There the *Explorer* was reassembled and launched by moonlight at high tide on 30 December 1857, becoming the fourth steamer to ply the river. She was only 54 feet long, including her 9-foot stern wheel, she had a 13-foot beam, and she drew fully 3 feet of water, leaving only a dangerous 6 inches of freeboard, when loaded. Her oversized boiler filled up a third of her open hull, but even then her engine was underpowered for its task—it had been ample for the Delaware but not for the Colorado. At her prow was a four-pound howitzer and behind the engine was a 7- by 8-foot cabin. Despite her strange appearance, Ives was quite proud of his boat, but the expedition's artist, Balduin Möllhausen, described her as a "water-borne wheelbarrow." The Cocopahs and Yumas also snickered at her size, dubbing her the "chiquito boat."[30]

On 31 December, with a blast of the steam whistle, Ives began his great adventure to determine the head of steam navigation on the Colorado, but that same day at Fort Yuma, George Johnson also set out for the same goal. Still smarting from having been cut out of the expedition he had promoted, Johnson had quietly decided to launch his own expedition unbeknown to Ives. He was aided in this endeavor by worsening federal relations with the Mormons, which spurred War Department interest in the possibility of bringing troops into Utah by way of the Colorado. Because of the sudden urgency of this question, the acting commander of Fort Yuma, ordered a detachment under Lt. James L. White to accompany Johnson on a speedy determination of the navigability of the upper river. Taking only twenty-five days

Ives's little steamboat, the *Explorer,* commanded by Captain Robinson, steaming upriver passed Chimney Peak in 1858. From a lithograph by expedition artist Balduin Möllhausen.

rations and, of course, a mountain howitzer, Johnson set out on the *General Jesup* with fifteen soldiers and a comparable number of armed civilians, including old trapper Paulino Weaver and Yuma chief Kae-as-no-com, known to the Yankees as Pascual.[31]

Johnson found the river above the fort quite similar to that below. For hundreds of miles it meandered through long valleys, its banks lined with cottonwoods, willows and mesquite, only occasionally interspersed with the corn, bean and mellon patches of the Mohave rancherias. Three narrow, but short, canyons separated the major valleys. The main difficulty in navigating was not rapids as some had feared, but the myriad of shallow sandbars which choked the channel. Since the river was then at its lowest stage, Johnson had a difficult time getting over some of the bars. This together with having to stop to forage for fuel made their progress much slower than that to which Johnson had become accustomed.[32]

Thus their provisions were nearly exhausted by the time they reached the first rapids on 21 January at the head of Pyramid Canyon, about eight miles above the later site of Davis Dam, and more than 300 miles above Fort Yuma. Though these rapids were by no means impassable, Johnson decided to turn back because of their short rations. First, however, he and White took a skiff a few miles on up into Cottonwood Valley to a place where they could see another forty miles of unobstructed river ahead. They concluded that they were within seventy-five miles of the mouth of the Virgin River and that, as Leroux had claimed, the Colorado was undoubtedly navigable to that point. They felt that they had shown that steamers could come within easy reach of the Mormon settlements, and they headed back triumphant. Though the distance to the Virgin was about twice what they estimated, their conclusion that it was the head of steam navigation was correct. It would still be many years, however, before this was demonstrated, so the question was still far from settled.[33]

On the first day of their return downriver, they had a surprise meeting. Stopping for wood at the head of Mohave Valley, they were startled to find a camel caravan. Lt. Edward F. Beale had brought the beasts to the American deserts to see if they would prove more useful than that old army standby, the mule. Beale was equally surprised and delighted to see a steamboat, for the ships of the desert refused to swim the river. Johnson agreed to ferry them across; it must have been a bizarre scene.[34]

Ives, in the meantime, had met with a distressing surprise of his own when he reached Fort Yuma and learned of Johnson's rival effort. The news cast an ill-humor over most of the members of his expedition. They were further embarrassed a few days later when, after much fanfare on leaving the fort, they ran hard aground on a sandbar in full view of their well-wishers. Their efforts to get the boat off provided several hours of amusement to the garrison and the Indians who gathered on the bank to watch. As it was nearly sundown when they finally freed the *Explorer,* they tied her up and made camp on the muddy bank, still in view of the fort but too embarrassed to seek its comforts for the night.[35]

In the days that followed, the overloaded craft, drawing even more than the *General Jesup,* ran aground many more times

A Mohave rancheria in the valley above the Needles sketched by Balduin Möllhausen with the Ives expedition.

despite the valiant efforts of Captain Robinson to find a deep channel through the maze of shifting bars. The Indians along the river found the groundings so amusing that Robinson soon came to recognize shoals simply by the excited crowds gathered at the bank in anticipation. The progress of the expedition was slowed virtually to a crawl. Crowded together on the little boat with no relief from sun, wind, sand or smoke, and limited to a ration of bread, bacon and sandy beans, the dour mood of most of the party grew even blacker and tempers flared at the slightest incident.[36]

A curious catharsis came on 30 January, however, when they met the *General Jesup* steaming home in triumph. The crews commiserated with each other, swapping anecdotes and tobacco, while Johnson told Ives and Robinson of the conditions upriver. Ives also took the opportunity to send one of his men back to the fort with Johnson to bring up more provisions by pack train. In his voluminous report, however, Ives made no mention of Johnson's achievement and only the most cryptic allusion to his expedition, for throughout it all, Ives held jealously to his epic view of the expedition as an achievement that would win him eternal fame.[37]

Johnson's expedition was not without its own frustrations, for just fifty miles above Fort Yuma, the *Jesup* struck a rock mid-stream and sank in three feet of water. Undaunted Johnson took the skiff to the fort and returned with the *Colorado*. Building a bulkhead around her sunken hull, he raised the *Jesup* in two days. Then, taking her down to the estuary, she was beached for repair. Lieutenant White dispatched his official report of the expedition to the War Department well before Ives ever reached the head of navigation. Johnson also further publicized his exploits with a lengthy letter in the San Francisco *Herald*.[38]

Ives, in the meantime, continued working the *Explorer* slowly up the river until 6 March when at the entrance to Black Canyon she, too, struck a rock. The jolt threw the men at the bow overboard, tossed Ives and others on the stern deck headfirst into the bottom of the boat and knocked loose the boiler and the wheelhouse. At first Ives feared the canyon had fallen in, but he happily found they were all still alive and afloat. Deciding not to tempt the fates further, he christened the boulder Explorer Rock and pronounced it the practical head of navigation. They had come almost five hundred miles from the mouth of the river and about forty miles above the point reached by Johnson.[39]

While they waited for the engineer to repair the steamer, Ives sent a party to survey a route to the Mormon road, and he and Robinson took a skiff to explore farther upriver. They went some thirty miles on through Black Canyon to the mouth of Vegas Wash which they mistook for the Virgin River. They returned all the more convinced that they should not attempt to go farther in the *Explorer,* though they, too, concluded that at higher water a steamer might make it all the way to the Virgin.[40]

Having completed the survey of the river, Ives's party moved back down to Beale's Crossing to await the pack train from the fort. While camped here they had a tense meeting with a Mormon "spy," Thales Haskell, who had been sent with Jacob Hamblin to investigate rumors that federal troops were headed up the river to invade Utah. Though he pretended to be an

This sketch of the *Explorer,* dwarfed by Mohave Canyon, gives more a sense of artist Möllhausen's apprehension than a view of the scenery.

immigrant bound for California, his true purpose was obvious to the members of the expedition. Moreover, he mistakenly concluded that the Ives pack train was the much feared "invading army." The Mormon scouts thus set out to find a place to make a stand against the invasion, while Ives began to fear that they would incite the Mohaves to attack him. There was much excitement, but it came to naught, for a settlement between the government and the Mormons was already under way.[41]

When the pack train finally arrived, Ives headed overland with most of the expedition to explore the country east to the Grand Canyon, while Robinson took the *Explorer* back to the fort. Though many of the party had complained continually while they were confined to the little boat, they were now sorry to leave her. Johnson purchased the *Explorer* at auction in July for $1,000, doubtless only to maintain his monopoly on the river, for the craft was too underpowered and unwieldy to have been of much use. Shipping the machinery to San Francisco, he used her for a time as a barge. While tied up to the bank at Pilot Knob during high water in 1864, she broke loose and drifted into a slough some sixty miles below. Subsequent changes in the channel left her high and dry. In 1929 a survey party discovered the old hull many miles from where the channel then lay. Her skeleton is still there, but her iron plate has long since been made into comales for baking tortillas.[42]

The Mohaves and Chemehuevis along the river had welcomed and befriended the steamboat expeditions, but the opening of the river to steam navigation marked the end of their isolation and their independence. Indeed, when Lieutenant Beale first met the steamer on the upper river, he lamented, "The steam whistle of the *General Jesup* sounded the death knell of the river races." Within a year the same steamboat that they had welcomed would return with soldiers to subdue them.[43]

In addition to bringing camels, Beale had opened a new wagon route to the West, running roughly along latitude 35 degrees north. He claimed it would soon become "the great emigrant road to California," surpassing all other routes. Instead it proved much more arduous and dangerous than any other. As with the camels, it proved to be an expensive failure. The first immigrant party to try the route turned back after nine were killed in an Indian attack at Beale's Crossing on the Colorado River in August 1858. When word of the attack reached Washington, the secretary of war ordered the immediate establishment of a military post at the crossing. The attack was actually made by a raiding party of Hualapais, joined by just a few Mohaves, but it was the Mohaves alone who would feel the wrath of the War Department.[44]

Lt. Col. William Hoffman was directed to establish the post and in January 1859 he arrived at the river with a small detachment to select a suitable site. The curious Mohaves gathering around his camp found his seemingly aimless maneuvers quite amusing. When a couple began mimicking his commands and movements, however, Hoffman became outraged at their lack of respect and opened fire on the crowd, killing nearly a dozen. He then beat a hasty retreat back across the desert to Los

Angeles. In defense of his actions he wrote, "If I had allowed this insolence to pass unnoticed they would have hooted us out of the valley."[45]

As a result of this encounter, Hoffman mounted a massive expedition to set up the fort and chastise the Mohaves. Five companies of infantry, 200 mules and 300 tons of provisions were sent to the river in February aboard the ocean steamship *Uncle Sam*—not to be confused with the river steamer. As if this force were not enough, they were joined by two more companies who had established an advance post, dubbed Fort Gaston, on the west bank of the river eighty miles above Fort Yuma. The expedition was a financial bonanza for Johnson who chartered both of his steamers for nearly three months transporting troops and supplies at $200 a day for the *Jesup* and $300 for the *Colorado*. The steamers carried the entire expedition from the mouth of the river to Fort Yuma. From there on the bulk of the troops and pack mules marched along the riverbank while the *Jesup* with two mountain howitzers and an artillery detachment accompanied them as a gunboat. The *Colorado* followed with additional supplies.[46]

The expedition reached Beale's Crossing 300 miles above Fort Yuma on 20 April 1859. There on a low bluff on the east bank of the river in what was then the New Mexico Territory, Hoffman established Fort Mohave. Hoffman was spoiling for a fight, but the Mohaves refused to be provoked. Johnson had privately expressed his confidence that the Mohaves were "not war-like" and would offer no resistance, but his personal profit from the business bought his public silence. Obviously a bit puzzled and overawed by so massive a show of strength, the Mohaves agreed to Hoffman's demand of hostages for both the attack on the immigrants and the fancied "threatened attack" on his own party in January. Cairook, their principal chief, even offered himself as one of the hostages and they were taken to Fort Yuma on the *General Jesup*. Hoffman won his demands so easily and bloodlessly that one of the newspaper correspondents accompanying him complained that the "war ended before it had really commenced."[47]

Two companies of the 6th Infantry and an artillery detachment were left under Maj. Lewis A. Armistead to build and garrison the fort. The rest of the expedition returned overland to the coast. Armistead, a Mexican war hero, was also dissatisfied with the bloodless conquest of the Mohaves, for he felt that not until they had been beaten on the battlefield would they show a proper respect for the army. His anxiety grew when Cairook was killed helping the other hostages escape, and, after some mules were stolen a few weeks later, he could control his martial zeal no longer. At dawn on 5 August he made a surprise attack on the rancheria of Cairook's successor, Iretaba, killing nearly two dozen Mohaves. Armistead's enthusiasm was not shared by his superiors, however, and he was promptly transferred. Capt. Dick Garnett was given command of the post and there was no further trouble between the troops and the Mohaves. Both Armistead and Garnett joined the Confederacy and were killed four years later at Gettysburg.[48]

The bustle of activity at the steamboat landing and cable ferry crossing at Fort Yuma is shown in this lithograph by George H. Baker.

Fort Mohave, established by Hoffman's expedition in April 1859, provided the first regular business for Johnson's steamers above Fort Yuma. Lithograph by George H. Baker of San Francisco.

Mohave chief Cairook, right, after befriending Ives's expedition in 1858, was taken hostage by Hoffman's expedition the following year and killed at Fort Yuma, whereupon he was succeeded by Iretaba, left. From a lithograph by Balduin Möllhausen.

Isaac Polhamus, Jr., who came to the Colorado in 1856, captained the steamers for nearly fifty years, becoming the dean of the riverboat men.

Life at Fort Mohave, like that at Fort Yuma, was not enviable duty. The soldiers' quarters were no more than large brush sheds, offering little protection from the heat or the elements. In addition, their rations, their clothing and even their shoes were in chronically short supply. For a time dysentery was killing two men a week.[49]

Carrying thirty to forty tons of freight on two feet of water, the *General Jesup* took nearly four weeks coming up from Fort Yuma. The lighter draft *Colorado* could make better time, but her new captain Isaac Polhamus, Jr., was very cautious in bringing her upriver. Polhamus, who later became the dean of Colorado steamboating, was not yet accustomed to its wild waters, having learned the trade on tamer streams like the Hudson and the Sacramento. On his first trip to Fort Mohave, in fact, Polhamus had refused to take the *Colorado* through Mojave Canyon, tying her up sixty miles below the fort, until Major Armistead sent down a detachment with orders to "bring her up or sink her." Most of the time, however, the *Colorado* was kept busy on the lower river carrying freight for both forts from the estuary to Fort Yuma. To improve shipping Johnson had a second stern-wheeler, the *Cocopah,* built in San Francisco for $35,000. Like the previous boats, she was shipped to the river in pieces and rebuilt at Gridiron Landing that summer. The *Cocopah* was the largest boat yet, measuring 140 feet in length by 29 feet at the beam, and drawing only 19 inches of water with 100 tons of freight. She was launched in August 1859, and David Robinson became her captain. With two swift stern-wheelers Johnson had no more use for the old side-wheeler *Jesup;* her machinery was sent back to San Francisco and her hull was floated down to its final rest in Minturn Slough.[50]

By the end of 1859 the steam navigation of the Colorado River had become routine, as Johnson's steamers regularly plied the river for a distance of 450 miles above its mouth, laden with troops and supplies for Fort Yuma and Fort Mohave. The establishment of Fort Yuma nine years before had prompted the opening of the river to steam navigation and with the establishment of Fort Mohave the conquest of the river was completed. Its exploitation now began.

This scene of the *Gila, Cocopah* and *Barge No. 3* lined up at Yuma waiting to discharge passengers and freight typified the booming river trade brought on by the mining rushes of the 1860s and '70s.

The Arizona Fleet

The discovery of rich silver lodes in Eldorado Canyon in the spring of 1861, followed soon after by the discovery of rich gold placers at Laguna de la Paz, drew wide attention to the riches of the country bordering the river and triggered the great Colorado River Rush of 1862. Thousands of miners, merchants and settlers poured into the country, providing the final thrust for the creation of the Arizona Territory the following year. Whereas just a few years before there had been only vague tales of gold on the Colorado, there were soon hundreds of producing mines and dozens of bustling camps and towns all along the river. The river boom was on, and the great mineral wealth of the region was opening up. What followed was a period of frenzied activity in the mines and camps along the river—one of unequalled profits for the river steamers and one of bitter rivalry for the river trade. This was the golden age for those slim stern-wheelers that came to be known as the Arizona Fleet.[1]

A brief rush had begun in the winter of 1858–59, simultaneously with the Mojave expedition, but it had quickly played out. In October 1858, Jacob Snively had found placer gold in the gulches bordering the Gila River just fifteen miles east of Fort Yuma. Within two months Gila City had sprung into existence. "Enterprising men hurried to the spot with barrels of whisky and billiard tables"; J. Ross Browne recounted, "Jews came with ready-made clothing and fancy wares; traders crowded in with wagonloads of pork and beans; and gamblers came with cards and monte-tables. There was everything in Gila City within a few months but a church and a jail." At its peak that spring it boasted a rough and tumble populace of some four hundred gold seekers and other "hard cases." George Johnson suddenly found himself "over head and heels in business," carrying men and provisions to the diggings. The activity even prompted formation of the Gila Mining and Transportation Company which sent a rival steamer

Johnny Moss, "the mining Kit Carson of the Coast," whose discovery of silver in Eldorado Canyon in 1861 precipitated the great mining rush to the Colorado, posed with Piute chief Tercherrum two years later for pioneer photographer R. d'Heureuse at Lower Camp in Eldorado Canyon.

to the river on the schooner *Arno*. This rivalry ended before it began, however, when the *Arno* capsized and was lost with all her cargo at the mouth of the Colorado on 17 March 1859.[2]

No sooner, too, was Gila City established, than the boom collapsed. Though there was still gold in the gulches, it was a mile or more from the river and the dirt did not pay enough to carry it down for washing. Dejected Yankee argonauts headed back to California, denouncing the diggings as a humbug, but their places were soon filled by Sonorans more accustomed to desert mining. "Dry washing" the dirt, they made good pay of five dollars a day or more. Their process was as simple as it was effective. They simply tossed the gold-bearing dirt in the air over a blanket, or a wide shallow bowl, much as one would winnow wheat, letting the wind carry away the lighter dirt, leaving the heavier gold behind. In this way the Sonorans worked the diggings for a few more years before they were finally worked out. Flood waters washed away most of the abandoned camp in 1862, and a later visitor found only "three chimneys and a coyote" to mark the spot.[3]

Limited as they were, the Gila diggings still gave an indication of the wealth the region held, and some of those who had left Gila City, such as Snively, stayed on the river to prospect further. Early in 1859 placer gold was also found at the Pot Holes on the bank of the Colorado eighteen miles above Fort Yuma, and by the end of that year gold had been discovered in bars on the river all the way up into Black Canyon, nearly a hundred miles above Fort Mohave. The placers at the Pot Holes, quite probably, had first been worked by the Spanish who briefly settled the ill-fated Mission San Pedro y San Pablo de Bicuñer near there in 1781. These discoveries attracted local attention, but it was not until 1862, following the startlingly rich strikes at Eldorado Canyon and Laguna de la Paz, that the great rush to the Colorado River began.[4]

The first of these discoveries was made by a prospecting party led by a buckskin-fringed former trapper, Johnny Moss. Moss had a gift of gab that could subdue a grizzly and an eye for mineral that earned him the title of "the mining Kit Carson of the Coast." He and two other old trappers, Paulino Weaver and Joseph Walker, played important roles in opening the mines all along the river. Moss's party led the way in April 1861 with the discovery of rich silver lodes in what came to be known as Eldorado Canyon on the west side of the river, some sixty-five miles above Fort Mohave. Ives's party had camped just above the mouth of Eldorado Canyon while they repaired the *Explorer* four years earlier, but they failed to discover its wealth. Moss located several claims, the richest of which proved to be the Techatticup and the Queen City. The Techatticup showed an eight-foot wide vein of silver chlorides, averaging $70 a ton at the surface and running as high as $3,666 in pockets at depth. News of the strike quickly reached Fort Mohave, enticing George Johnson to bring the *Cocopah* up to the mouth of the canyon that same month. Eager for new business he promised to land freight from San Francisco at five cents a pound, much less than the twelve cents a pound overland freighting rate across the Mojave Desert.[5]

Having located the best ground, Moss headed for the coast to spread word of his new El Dorado and find buyers for his claims. He succeeded at both. That fall the rush to Eldorado

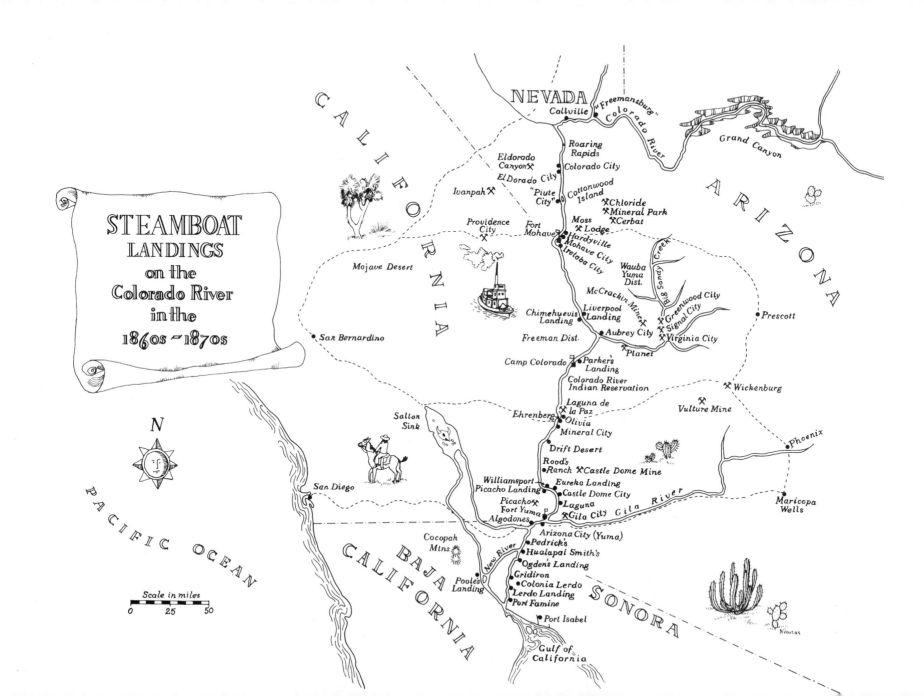

Canyon began. Soon over seven hundred mining claims had been located, blanketing the canyon like a crazy quilt. Levi Parsons, George Hearst and other prominent San Francisco capitalists bought up claims and floated dozens of million-dollar companies. Some worked the mines; others worked the stockholders. Unlike placer mines, where a man with little to his name but a shovel and a blanket could make good money if the ground were rich, lode mines generally required considerable capital to open them, hoist out the rock, and mill or smelt the ore.[6]

Several camps sprang up in the canyon and along the Colorado near its mouth. Each vied to be the "boss camp" of the district. San Juan, or Upper Camp, at the head of the canyon was ten miles from the river; Alturas and Louisville were halfway down the canyon near the Techatticup mine; and El Dorado and Colorado cities were down at the mouth of the canyon on the riverbank. Each of these burgeoning camps boasted several stone cabins and several dozen tents, but they offered few amenities. Their barren, sunburnt surroundings, in fact, prompted one newcomer to complain that the country looked "like hell with the fires put out." But it was silver, not scenery or comfort, that drew men there by the hundreds.[7]

At first only the richest high-grade ore was worked and shipped out for milling. It had to run a couple hundred dollars a ton to pay much profit above the combined costs of mining, milling and shipping to San Francisco. Thus in the fall of 1862, former Los Angeles state senator Col. J. R. Vineyard began accumulating the machinery for a custom stamp mill on the bank of the river to cut total operating costs in half by eliminating shipping. It was finally completed late in 1863, becoming the first stamp mill in what was then Arizona Territory. Though it was a makeshift affair, pieced together out of discarded equipment from the Mother Lode, it did a thriving business on Hearst's Queen City ore and chlorides from several of the smaller mines. Its success prompted the erection of a rival, the New Era mill, which first dropped stamps on Techatticup ore in April 1865. The mills' avaricious consumption of wood, however, rivaled that of Johnson's steamers. To supply the demand an enterprising Captain L. C. Wilburn started a "line of first class barges," the *El Dorado, Veagas* and *Colorado,* to bring driftwood up from Cottonwood Island by a combination of sail and "Mohave-Piute power." He also poled his barges on up the Colorado to the mouth of the Virgin River to bring down salt for the mills.

With the loss of downriver ore freight, once milling was begun at Eldorado Canyon, Johnson made less frequent trips above Fort Mohave and he raised the freight rates to help offset the loss. This was still the most difficult stretch of the navigable river, and for nearly half the year from November to April the water was usually too low over the bars and rapids to float enough cargo to make the trip pay. Even at high water during the spring floods it took several days to make the sixty-five-mile trip up from Fort Mohave, but coming back down under full steam Johnson could make it in lightning time of *four* hours![8]

Right at the height of the frenzied rush to Eldorado Canyon came news of fabulous new placers farther down the Colorado.

This compounded the excitement and turned the rush into a stampede. In January 1862, while trapping on the Colorado 130 miles above Fort Yuma, Paulino Weaver discovered traces of gold in an arroyo on the east side of the river, in what became known as the La Paz diggings. Weaver showed the gold to José Redondo who was working a crew of Sonoran *gambusinos,* or miners, at Gila City. He took a few men up to the site, and while dry washing the first pan of dirt from the neighboring gulch he found a two ounce *chispa,* or nugget. As others learned of the strike the rush began. After that new discoveries were made almost daily in every gulch and ravine for twenty miles to the south and east.[9]

Within a year there were nearly two thousand men at the diggings—Sonorans, Papagos and Yaquis from Mexico, die-hard forty-niners from the Mother Lode, greenhorns from San Francisco, cowboys from the ranchos and, of course, Mohaves and Yumas from the neighboring valleys. The Sonorans and Yaquis outnumbered the rest nearly four to one. Every gulch was crowded with a couple hundred men; a few made $100, or even $1,000, in one day, while others could not make a cent. Most seemed, however, to average about $4 a day—good wages at the time. Each gulch also had its own camp. Campo de Ferrar, where Juan Ferrar found a magnificent four-and-one-half-pound nugget, was the richest; the other camps—Los Minos Primeras, Campo en Medio, American Camp, Los Chollos, Las Pozas and La Plomosa—each hoped to rival it. All of these camps, however, suffered from lack of water, which had to be packed eight miles or more from the river and sold for three or four bits a gallon. The boss camp of the district thus became Laguna de la Paz on the Colorado bottomland just west of the diggings.[10]

La Paz grew quickly and chaotically from a few brush shacks and a handful of inhabitants to a bustling town of rambling adobes teeming with nearly a thousand fortune seekers. "Everything about the place was in shiftless confusion," one visitor noted, "and if laid out properly on a map, the town would have resembled, as nearly as anything, a cobweb drawn and blotted by a young child on a rumpled piece of paper." Nonetheless when the Arizona Territory was created in 1863, La Paz was its largest and "most flourishing" town. Though it missed being named the capital of the territory, it was made the seat of the newly created Yuma County. The lure of aguardiente and a game of monte in La Paz's cantinas drew the gambusinos down from the diggings in the evenings, while fandangos and an occasional bullfight enlivened the Sundays. La Paz also attracted all the roughs who followed the rush and they regularly had their "man for breakfast." In the diggings themselves, heat and lack of water also took their toll.[11]

La Paz had one further drawback—it soon found itself some two miles from the riverbank. Ironically when the town was settled in the spring of 1862 it was laid out right on the riverbank, but the river was at a record flood peak that spring and when the water began to fall the steamers could no longer reach the landing. Thus a rival town was laid out at the river the following year to serve as the steamer landing and to become, its founders hoped, the "commercial emporium" of the district. It was named Olivia, or later Olive City, after Olive Oatman, a young girl whose exploits as a captive of the Mohaves had become a best seller. Olivia was a strictly "White" town—no Negro, Indian or Chinese being allowed to buy a lot—and it became the retreat for secessionists and others whose bigotry made them uneasy in more cosmopolitan La Paz. In March 1863, even the mines became segregated when the miners at Olivia seceded from the La Paz Mining District, forming the Weaver district in an attempt to exclude Mexicans and Indians from the mines. Olivia and the La Paz diggings, in general, also served as a staging point for Confederate sympathizers heading east to join in the Civil War. To break up this activity a detachment of Union infantry from Fort Mohave set up camp halfway between La Paz and Olivia in September 1863. That same fall another rival, Mineral City, was laid out at Bradshaw's Ferry, a mile below Olivia. Neither Olivia nor Mineral City, however, grew to much more than a dozen adobes and jacals before they were eclipsed by yet another rival, Ehrenberg, established halfway between them. Ehrenberg's success was assured by the fact that two of its organizers were Johnson's steamer captains, Isaac Polhamus and David Robinson, who naturally saw to it that all the freight for the region was landed there, rather than at its rivals. As the placers gave out Ehrenberg eventually surpassed even La Paz as the chief town in the district. By the late 1860s most of the La Paz merchants, led by the Goldwater Brothers, had relocated there and it had become the main shipping point for mines of central Arizona.[12]

Johnson's steamers did a bonanza business carrying men and supplies to all the new mines opened up along the river and in the interior. He secured a virtual monopoly on the trade, for even though two new wagon roads were opened across the desert from Los Angeles, the overland teamsters demanded $250 a ton, while Johnson could land freight from San Francisco for less than half that price.[13]

From Eldorado Canyon and La Paz prospectors had soon fanned out into the surrounding hills in search of new bonanzas. New claims were located and new districts were formed in rapid succession. By the end of 1863 there were more than a dozen mining districts blanketing an area the size of New England, stretching for thirty miles on either side of the river all the way from Fort Yuma to Eldorado Canyon.

Again Johnny Moss led the way. Having profitably disposed of his claims in Eldorado Canyon, Moss organized a new party to prospect the hills on the opposite side of the river. In the fall of 1862, just nine miles northeast of Fort Mohave he discovered a rich surface cropping of gold-bearing quartz which he modestly named the Moss Lode. He unblushingly proclaimed his new discovery "the most immense thing on record," claiming it surpassed even the great Comstock Lode with ore assaying as much as $24,000 a ton and running for more than three miles with an average width of 100 feet. Other only slightly less promising claims were found nearby and the San Francisco Mining District was born. Unloading this wonderful new ledge on San Francisco capitalists, Moss once again headed back to the Colorado, where

Ehrenberg, founded in 1866, quickly became the main upriver port supplying the various mines and settlements of central Arizona, doubtless because two of Johnson's steamer captains, Polhamus and Robinson, were among its promoters.

he made new strikes in the Hualapai Mountains to the east and organized the Wauba Yuma district.[14]

Others prospecting the west side of the river to the south of Eldorado Canyon found silver in the Providence Mountains near the Government Road from San Bernardino to Fort Mohave. There the Rock Springs district was formed in May 1863 and the Macedonia district the following year. A string of stone cabins and tents, known as Providence City, flourished briefly near the mines. Across the river opposite Eldorado Canyon the Pyramid district was opened. Farther south soldiers from Fort Mohave found copper ore in the Mohave Mountains, five miles west of the river in what they called the Iretaba district in honor of the Mohave chief. In September 1863 the "blue coats" made another strike, organizing the Sacramento district in the Cerbat Range just north of Moss's Wauba Yuma.[15]

As the mines around Fort Mohave were opened, new river ports sprang up to handle their trade. Iretaba City, laid out in January 1864, just two miles below the fort, was the first. But it was eclipsed two months later when William Harrison Hardy established a ferry and laid out a rival town nine miles to the north. Hardyville was much closer to the mines, and according to its founder it was at the "practical head of navigation," being accessible by steamer even during low water. Hardy, backed by George Johnson to the extent of $40,000 in shipping credit, cornered the trade on the upper river and for a time became the principal shipping agent for Prescott and much of the interior. During the low water season from November to April, when

William Harrison Hardy, backed by Johnson, laid out the town of Hardyville as a supply point for the upriver mines in 1864 at what was the "practical head of navigation" for steamers at low water.

Johnson refused to try taking his boats above Hardyville, it also became the transfer point for freight to Eldorado Canyon and points upriver. Captain L. C. Wilburn, the bargeman, handled the shipping during this season, floating goods from Hardyville to Eldorado Canyon for forty dollars a ton. A third port, Mohave City, adjacent to the fort, cornered a portion of the shipping trade until the fort commander expanded the post boundaries in 1869 and ordered the residents out.[16]

Prospectors from La Paz also made new strikes on down the river. In the fall of 1862 more than three hundred miners rushed to new placers discovered on the California side of the river, fifty miles south of La Paz, in the Picacho district. Across the river more gold was found and Laguna camp sprang into existence. Farther back from the river silver-lead ore was found in the Castle Dome Mountains, which gave their name to another new district organized the following spring. A year later the Eureka district was formed when new silver strikes were made in the nearby Chocolate Mountains. Castle Dome City and Williamsport quickly grew up on the riverbank to serve the new districts.[17]

The Williams Fork district, just north of La Paz, was also organized in 1863 after Richard Ryland discovered the Planet lode a dozen miles east of the river. The Planet ore, running forty percent copper, was so rich that it paid a handsome profit even after being shipped halfway around the world to Swansea, Wales, for smelting. Other copper mines were opened nearby, and primitive smelting furnaces were set up both at the camp of Planet and at Aubrey City, the river landing for the district. Copper strikes that same year in the mountains bordering the river on the west also led to the forming of the Freeman, Marengo and Chimawavo districts.[18]

At the same time that these discoveries were being made in the hills along the river, old trappers Paulino Weaver and Joe Walker led prospecting parties to new placers and lodes more than a hundred miles to the east. The ensuing rush to these diggings opened up many more mines in the interior. The richest of these was the famous Vulture mine in the Hassayampa district discovered by Henry Wickenburg late in 1863. Remote as these mines were, they, too, were supplied primarily from the river and its steamers, further swelling the river trade.[19]

The Colorado River rush suddenly made George Johnson and his two partners wealthy men. Ben Hartshorne who handled the business end of the partnership in San Francisco invested his earnings in the California Steam Navigation Company, which monopolized the lucrative trade on the Sacramento River. He became president of that company. Captain Alfred Wilcox settled down in San Diego where he opened a bank and bought a fine house and a small yacht, the *Restless*. Johnson, still in his mid-thirties, gradually turned over the day-to-day management of affairs on the river to Isaac Polhamus, so that he, too, could indulge in the good life. In June 1859 he married Estefana Alvarado, a young niece of former California Governor Pio Pico, and bought the old Rancho de los Peñasquitos near San Diego together with several thousand head of cattle. There between occasional trips to the river he lived the life of a California Don, cultivating expensive tastes. Johnson also dabbled in politics, winning San Diego's seat in the state assembly in 1862 on the solid vote of 250 soldiers stationed at Fort Yuma.[20]

The steamboat business, however, suffered from Johnson's inattention. Preoccupied with his ranch and politics, he failed to expand his fleet to meet the increased demands of the rapidly developing mines along the river. Though he had built a new boat in the spring of 1862, it was only to replace the aging *Colorado*, whose hull was beginning to show the wear of six years of dragging the sandbars and buffeting the tidal bore. The new boat, also named the *Colorado* (II), had the former's machinery and was only slightly larger, 145 feet long with a 29-foot beam. Fearful that she might be captured by Confederate sympathizers, Johnson had her built at Arizona City under the guns of Fort Yuma rather than at the estuary as before.[21]

Thus despite the establishment of Fort Mohave and the opening of a myriad of mines along the river, Johnson still had only two boats on the river—the same as he had had since 1856 when he was supplying only a handful of troops at Fort Yuma. For the few months of high water in May and June the *Cocopah* could deliver nearly four hundred tons of freight a month to the La Paz landings—making a round trip from Fort Yuma in about four days with sixty tons of cargo. This rate, however, quickly decreased as the water in the river fell and the trips became much slower; by December she could barely make two trips a month with forty tons—carrying an average of only eighty tons a month. The *Colorado*, working the river from Fort Yuma to the estuary, did little better. The two boats were simply unable to carry the ever increasing volume of freight shipped to the river.[22]

By the fall of 1863 there were more than twelve hundred tons of freight piled up at Arizona City and at the mouth of the river waiting to be brought upriver, and at nearly every landing there were ten to fifty tons of ore waiting to be brought down. Some of the freight had already been lying there for a couple of months. Most of it was destined to remain for several more months until the spring rise of the river. Merchants upriver rapidly ran out of goods and prices rose astronomically. Moreover much of the freight that was brought up was brought on consignment to the steamer captains and officers to reap quick profits from the inflated prices, leaving behind identical shipments consigned to the merchants. Even these shipments, however, ended in November when a fall in the river left the *Cocopah* "high and dry" on a sandbar thirty miles above La Paz.[23]

Not only was the river freighting slow but the rates were considered excessive compared to those on other western rivers. La Paz merchants particularly felt discriminated against, since Johnson charged them seventy-five dollars a ton to carry goods 280 miles upriver to La Paz, while he charged only twenty-five dollars a ton to bring goods 200 miles to Williamsport.[24]

On 1 December 1863 the merchants and miners at La Paz held a heated protest meeting, condemning the Johnson company as a "swindling monopoly," and accusing them of trying to "freeze out" the miners in order to gain control of the mines. To

The steamer *Colorado* (II) making a downriver stop under a dramatically threatening sky at Williamsport opposite Picacho or Chimney Peak.

seek relief they voted to send a representative, Samuel Adams, to San Francisco with a petition calling for the establishment of an opposition steamboat line on the Colorado River. Adams, a thirty-year-old, fast-talking Pennsylvania lawyer, had come to Arizona only a few months before, seeking an office in the newly created territorial government. To make a name for himself in the hopes of winning a seat in Congress, he quickly became the most vocal critic of the Johnson company. For his efforts he soon earned the irreverent nickname of "Steamboat Adams," but he ultimately failed to win the congressional seat.[25]

In San Francisco, however, Adams met with swift success. Rhapsodizing over the profits to be made on the river trade, he won the Chamber of Commerce's enthusiastic endorsement of a rival line. Then with a subscription of $25,000 in freight money raised from interested merchants, he persuaded Captain Thomas Trueworthy, who was planning to put a steamer on the Yang Tse River in China, to take it to the Colorado instead. Within a month the opposition Union Line was born. In February 1864 Trueworthy sent Captain Charles C. Overman to the Colorado with his steamer, the *Esmeralda,* and a barge, the *Victoria,* converted to a four-masted schooner. The *Esmeralda* was a 93- by 20-foot stern-wheeler of 33-inch draught, built two years earlier for the Sacramento River trade. Smaller than Johnson's steamers, she had a capacity of only fifty tons, but she had a more powerful engine and could easily tow a barge of 100 tons. The vessels reached the Colorado in March. Overman set up the *Victoria* as a store ship at the mouth of the river, but she was soon broken up

Thomas E. Trueworthy started an opposition steamboat line on the Colorado in 1864 in an unsuccessful bid to break Johnson's monopoly of the river trade. Lithograph by Clorinda.

by the tidal bore and was towed up to Port Famine Slough, where she later burned. In the meantime Overman constructed a 128-by 28-foot barge, the *Black Crook,* for use on the river.[26]

In early May 1864 Trueworthy arrived to take the steamer upriver with his first cargo. Even with the barge in tow, he claimed a record time of three days, eight hours on the 150-mile run from the estuary to Fort Yuma. Trueworthy later built a second barge, the *Pumpkin Seed,* but it swamped and sank with a load of iron in 1867. Though Johnson had previously built a 150-ton barge for use at tidewater, Trueworthy was the first to take a barge in tow upriver. So as not to hinder the maneuverability of the steamer, the barges were towed on a 100-foot cable attached to a short mast amidship on the steamer, rather than to the stern. Each barge had its own helmsman whose job it was to keep it exactly in the wake of the steamer. The system, long used on other rivers, was so successful that Johnson built two more barges for the same purpose the next year—naming them imaginatively *No. 2* and *No. 3*—and later followed with *No. 4*.[27]

Adams had unwittingly been so persuasive in his call for opposition steamers that a second rival entered the field just a few months later. Alphonzo F. Tilden, managing director of the Philadelphia Silver and Copper Mining Company, enticed both by the potential freighting profits and the need to ship ore from their mines in Eldorado Canyon and the Freeman District opposite Bill Williams' Fork, decided to put a steamer of his own on the river. Sparing no expense, he had a swift, shallow-draught sternwheeler, the *Nina Tilden,* built by Martin Vice in San Francisco. She was 97 feet long, with a 22-foot beam, a draught of 12 inches light, and a capacity of 120 tons. She was launched 23 July 1864, and on her trial run on the bay she averaged a speedy sixteen knots. Her captain, George B. Gorman, a veteran of the Sacramento and Fraser rivers, sailed her down the coast to the Colorado the following month. With a barge, the *White Fawn,* brought down on the schooner *Sarah,* the *Nina Tilden* entered the competition for the river trade in September 1864. To supplement his own ore shipments, Tilden advertised to buy sacked ore in lots of ten tons or more at every landing on the river.[28]

In the meantime, George Johnson, finally spurred to action by the challenge of competition, had built a third boat for his own fleet. This new steamer, the *Mohave,* launched at the estuary about the end of May 1864, was 135 feet long with a 28-foot beam. In July, under Captain Polhamus, she joined the *Cocopah* on regular runs above Fort Yuma and soon set a record time of ten days and two hours for the 365-mile run to Eldorado Canyon, putting off 225 tons of freight at the landings along the way.[29]

The number of steamers working the Colorado had suddenly leaped from just two in the spring of 1864 to five by the fall—from a damnable deficiency to a damnable excess. Whereas before, working constantly, the steamers could not move half the freight sent to the river, there was not now enough to keep all of the steamers running half the time, even though the amount of freight was ever increasing. Together, the addition of new steamers and the introduction of barges had increased the capacity of the river fleet nearly fivefold.

The towing of barges to increase the freighting capacity of the steamboats was introduced by Trueworthy and quickly adopted by Johnson, whose *Barge No. 4* is being towed through Red Rock Gate above Picacho.

The *Mohave* (I) was built in 1864 and the *Colorado* (II), behind with only her stack and pilot house visible, was built two years earlier to help carry the flood of freight pouring into the newly opened mines upriver from Yuma.

With the advent of the rival steamers a bitter war ensued over the freight already piled up at the estuary. Johnson moved much of it just a few miles upriver to the Gridiron, thereby getting a lien on it so his competitors could not take it. To assuage La Paz merchants, however, he cut shipping charges to that landing by nearly half—to $40 a ton. The rival boats thus had to rely on agents in San Francisco to ship freight through them. Even when they did get a cargo they found the trip upriver slower going than expected, since they had to stop to gather their own fuel—Johnson had bought out most of the woodyards. Adams accused Johnson and his men of even more sinister harassment against the *Esmeralda*—gumming her machinery, setting fire to her cabin, cutting her loose from her moorings and floating timbers downriver to wreck her. To make matters worse Trueworthy complained he could not even get insurance on his boat and cargo because of Johnson's pressure on the brokers.[30]

In the fall of 1864 both Johnson and his competitors, lacking sufficient work to keep their boats busy, made a bid for part of the lucrative trade with Utah. Freighting over the Mormon road from Los Angeles to Salt Lake City was then a $3 million a year business. Johnson offered to carry Utah freight for only $65 a ton upriver as far as Eldorado Canyon during high water and to Hardyville at low water. This would cut the overland haul by nearly half and the freighting rate by a third, saving Salt Lake merchants more than $100 a ton. Not to be outdone, Trueworthy, egged on by Sam Adams, offered to land freight year-round at the mouth of Vegas Wash, forty miles upriver from Eldorado Canyon. The Deseret Mercantile Association eagerly accepted both offers and Brigham Young, who for years had longed for a sea route to Zion, quickly dispatched Anson Call to select the site for a landing and warehouse at the river. On 17 December 1864 he settled upon a spot, which he modestly christened Callville, on a gravelly point some six miles above Vegas Wash. There he built a large stone warehouse and corral. A year later the Arizona legislature created Pah-Ute County out of that portion of the territory along the Colorado north of Roaring Rapids and Callville was made the county seat. The honor was short-lived, however, because Callville and most of the county were ceded to Nevada in May 1866.[31]

The race to land the first Utah freight began as soon as Call headed for the river. Because it was low water by then, Johnson refused to take freight above Hardyville, but William Hardy offered to float it the remaining ninety miles to Callville by barge. Building a 25-ton, 50- by 8-foot barge, the *Arizona,* he set out with his first load of freight on 2 January 1865. Using poles, oars and even sail on two favorable days, he reached Callville in twelve days. At the same time Trueworthy and Adams headed upriver from La Paz in the *Esmeralda* with a bargeload of lumber and merchandise, intent on proving their claim that steamers could reach Callville any time of the year. They got only as far as Roaring Rapids, however, twenty-eight miles short of their destination. They could have gone all the way, Adams claimed, but turned back on learning that Call had returned to Utah after hearing that the steamer had broken down.[32]

Undaunted Trueworthy and Adams tied up the *Esmeralda* at Eldorado Canyon and headed overland for Salt Lake to proclaim their "complete success" in opening a new "era in the trade of Utah." Trueworthy had, in fact, succeeded in doing very little. He had taken the *Esmeralda* only seven miles beyond the point reached by the *Explorer;* he had not succeeded in winning a profitable piece of the Utah trade even away from Johnson and Hardy; and he was heavily in debt. Thus after finding a buyer for the freight left at Eldorado Canyon, Trueworthy hurried back to San Francisco to get additional financial backing for the venture. There, after several months of hard talking, he succeeded in rounding up new backers. In the summer of 1865 they bought both the *Esmeralda* and the *Nina Tilden* and organized the Pacific and Colorado Steam Navigation Company with a capital of $200,000. The officers of the new company were a hardware manufacturer, Joseph W. Stow, president, and a wholesale grocer, Kimball C. Eldredge, secretary. Adams on the other hand had cooled briefly on the Colorado River business and had gone to the Columbia River to look into the possibility of supplying Salt Lake via that artery.[33]

In the meantime, George Johnson had made an attempt of his own to open more direct trade with Utah but was thwarted again by low water. The river was so low that summer that the *Cocopah* was not able to make it up to Eldorado Canyon; therefore the Utah freight was again left at Hardyville to be poled upriver by barge. Moreover, as the river continued to fall the *Cocopah* was soon unable even to reach Hardyville and she ended

Anson Call was sent to the Colorado by Brigham Young in 1864 to find a suitable landing for Utah freight in the hopes of opening trade via the river. Photographed from a steel engraving.

up stranded on a bar just above La Paz for six weeks.[34]

After an unsuccessful attempt to underbid Johnson for the government supply contract, Trueworthy once again tried to open trade with Utah. With high water in the summer of 1866 the *Esmeralda* headed upriver for Callville with a barge and ninety tons of freight. She was commanded by Trueworthy's first mate, Captain Robert T. Rogers, and accompanied, of course, by the ubiquitous Sam Adams. This time she succeeded, after a fashion. On 8 October 1866, after a struggle of three months, the *Esmeralda* edged into the landing at Callville, finally opening what one wag called "the backdoor of Utah." Her principal difficulty had been lack of firewood which meant insufficient power to stem the main current. When her engine proved inadequate at Roaring Rapid in Black Canyon, a ringbolt had had to be set in the canyon wall and a line run to her capstan to pull her through. Nonetheless, she had gotten through and Callville was triumphantly proclaimed the new head of steam navigation on the Colorado.[35]

Once again it was a hollow triumph, for the Pacific and Colorado Steam Navigation Company, still unable to gain a profitable piece of the river business, was deeply in debt. No sooner had the triumphant *Esmeralda* returned downriver for her second load of cargo, than she and the *Nina Tilden* were seized by the sheriff. Both boats remained tied up at Fort Yuma for more than a year while their creditors tried to float a new firm, the Arizona Navigation Company. When this effort was finally abandoned in the fall of 1867, Johnson bought the boats and the rivalry for the river trade came to an end.[36]

The competition for the river business had greatly benefited miners, merchants and settlers throughout western Arizona. It had forced Johnson both to lower shipping rates and to improve service with the addition of new steamers and the introduction of barges. The further development of the territory that this, in turn, stimulated brought further trade and even greater profits to Johnson. The only losers were Trueworthy, Adams and their backers; their only reward was a resolution of thanks from the Arizona legislature praising their "untiring energy and indomitable enterprise." Since Johnson's hold on the bulk of the river business, especially the Fort Yuma and Fort Mohave supply contracts, had never been seriously threatened by his rivals, the only real chance for a successful opposition line had been in the opening of new trade with Utah via the Colorado. Even a lingering hope of this was quashed in May 1869 with the completion of the Central and Union Pacific railroads.[37]

By then Callville's only inhabitants were a gang of fleeing horse thieves who wrenched the door off the abandoned warehouse to raft down the river. But Adams, always an aberration on the course of destiny, set out undaunted for the Rocky Mountains to demonstrate that the Colorado was, in fact, navigable clear to its headwaters and was still destined to become the main artery of Utah trade. Two years before, a tattered prospector, James White, had been dragged from the river at Callville after a harrowing voyage through the Grand Canyon on a few logs, but Adams was determined to try to do it in style. When this effort ended in debaucle, he finally returned to the East to end his years

The stone warehouse built at Callville only received one shipment of freight on Trueworthy's *Esmeralda* in 1866 before the idea of trading with Utah by way of the river was abandoned. This photograph of the warehouse and corral was taken in the 1930s just before they were flooded by Lake Mead.

trying to exact money from the Congress for his unsolicited adventures.[38]

Johnson faced no further threat to his monopoly of the river trade for a full decade; his business thrived with the continued growth and development of the territory. By the late 1860s Johnson's boats were carrying roughly six thousand tons a year and earning a quarter of a million dollars annually. On 20 December 1869, Johnson and his two partners, Hartshorne and Wilcox, officially incorporated as the Colorado Steam Navigation Company, issuing $500,000 in stock and taking in Edward Norton, a former California Supreme Court judge, and Richard D. Chandler, a San Francisco coal dealer, as partners.[39]

By then the company was running four steamers, the *Colorado* (II), the *Mohave,* the *Nina Tilden* and the new *Cocopah* (II). The latter, seven and a half feet longer and a foot slimmer than her namesake, was built in March 1867. In addition they had half a dozen barges, the *Black Crook, White Fawn, Yuma, No. 1, No. 2,* and *No.3.* Their payroll listed nearly a hundred men whose jobs were manning the boats and handling the freight. As general superintendent, Captain Isaac Polhamus oversaw all operations, while Johnson once again luxuriated at his rancho. A new generation of men now captained the steamers—Alexander D. Johnston, John Alexander Mellon, Charles Caroll Overman, William Poole and Stephen Thorne. Jack Mellon was destined to become the most famous of the Colorado River captains, working the river for over forty years. He was a New Brunswick sailor, who first went to sea at the age of ten. He had come to the river in 1864 with Overman and Thorne to work for Trueworthy. The steamboat captains earned $200 a month—good pay at the time. Their crews consisted of an engineer, who also drew from $160 to $200 a month, a first mate, a fireman and a cook, each of whom made $75, an assistant fireman who earned $40, a steward who was paid $25, and six to seven deck hands, whose wages depended on their nativity. Yankee deckhands got $35 a month, Sonorans and Kanakas $25, and Cocopahs, Yumas, and Mohaves $15. The barges also had a captain, or steerer, who drew $75 a month, and two deckhands. All hands received board.[40]

At the mouth of the river the steamboatmen had finally located a safe harbor, Port Isabel, for transferring passengers and freight. It was a slough which opened into the gulf a few miles east of the mouth of the Colorado. Sam Adams claimed that he and Trueworthy had discovered the slough in 1864, but it took its name from the schooner *Isabel*—captained by William H. Pierson—which first ventured into the slough in the spring of 1865. Three miles up the slough from the gulf the tidal bore was much less severe than on the Colorado, so seagoing vessels could anchor in relative safety while they transferred their freight to the barges. In 1867 the barge *White Fawn* was lengthened and moored in the slough as a wharf boat. Three miles farther up, where the slough widened, the company diked some twelve to fifteen acres for a shipyard, complete with dry dock. Quite a settlement grew up there on Mexican soil. Strung out in a single row were the carpenter shed, blacksmith shop, company office, cookhouse, mess hall, meat house, storeroom, a couple cottages

The *Cocopah* (II), built in 1867, completed Johnson's fleet of three steamers operating in the late 1860s. She was finally retired in 1879 after the Southern Pacific railroad bridged the river.

and the old steamer *Cocopah,* hauled out on the bank as a lodging house for the workmen. Behind these stood a henhouse, hogpen and water tank. When a steamer was in the dock for repair, the place buzzed with activity as the carpenters, caulkers, machinists, boilermakers, coppersmiths and painters swarmed over the vessel.[41]

Captain Robinson, originally a ship's carpenter himself, took charge of the shipyard and freight transfer at Port Isabel, while still hunting for Raousett de Boulbon's treasure in his spare time. Life at Port Isabel, however, proved to be a bleak and lonely existence for his young bride and baby. Thus in 1873, after twenty years on the river, Robinson finally left the Colorado to run a pleasure boat on Clear Lake north of San Francisco. There on 17 July 1874 he died suddenly of a lung hemorrhage at the age of forty-nine.[42]

Port Isabel became the final resting place of the *Nina Tilden* in 1874 and the *Mohave* the following year, but the *Nina Tilden* did not rest long. Leaking badly, she broke her lines during high tide, got caught broadside and rolled over, bottom up, blocking the channel till she was chopped up. The *Mohave* was hauled out next to the *Cocopah,* and her machinery was sent back to San Francisco, where it was put on the Sacramento River steamer *Onward.*[43]

With the dry dock it was much easier to keep the boats in repair, so during the ten years existence of the shipyard at Port Isabel only two boats were actually built there. Both, however, vied for the title of "Queen of the Fleet." The first was the 236-ton stern-wheeler *Gila* completed by Captain Robinson in January 1873. She was 149 feet long, with a 31-foot beam, a depth of 3.5 feet, and drew only 16.5 inches of water. She proved to be the most durable boat ever put on the Colorado. Working the river for more than a quarter of a century before she was rebuilt as the *Cochan* to run for another decade, she stopped only when the river was finally closed to steam navigation. Her rival was the *Mohave* (II), completed in February 1876—the only double-stack steamer ever put on the river. In outward appearance she resembled the palatial riverboats of the Mississippi, but her interior furnishings were much less posh. She was in great demand as the grand excursion boat of the Colorado, but like the *Gila,* she was primarily a workboat and ran for almost as many years as her rival. She was half a foot longer, half a foot broader and half a foot shallower than the *Gila,* and she drew only a foot of water which permitted her to venture farther up the river and farther into the sloughs than any other boat. With Captain Jack Mellon at the helm she eventually made twenty trips to the mouth of the Virgin River nearly six hundred miles up the Colorado—truly establishing it as the head of steam navigation.[44]

The greatest growth of the Colorado Steam Navigation Company came in 1871 when they purchased an ocean-going steamer, the *Newbern,* and opened a direct steamer route from San Francisco to connect with the riverboats at Port Isabel. The *Newbern,* built in Brooklyn in 1852, was a 943-ton, 375-horsepower, propeller-driven vessel, 198 feet long with a 29-foot beam. Under Captain A. N. McDonough she began regular

Captain Robinson on the main deck of the *Colorado* (II) supervising her overhauling in the dry dock at Port Isabel in 1868.

The *Gila*, launched in 1873, was the first steamboat built at Port Isabel and the most durable ever put on the Colorado. She ran for more than a quarter century.

The *Mohave* (II), launched at Port Isabel in 1876, was the largest steamboat ever run on the Colorado and the only double-stacker. She was also the most popular boat for excursions. Here she is decked out for children on a May Day school picnic in 1876.

monthly trips to the river on 2 July 1871. She made the 2,100-mile voyage in just twelve days, nearly cutting in half the usual sailing time and greatly expediting passenger and freight service. The all-steam service proved so successful, in fact, that the company bought a second steamer, the *Montana,* two years later, offering sailings every twenty days. The *Montana,* a 1,004-ton side-wheeler, built in Bath, Maine, in 1865, began operation on 25 October 1873, under Captain William Metzger. She was, however, an ill-fated ship. In December 1874 she ran aground in the Gulf of California and had to be towed back to San Francisco for repairs. Three months later she was back in service, but on 14 December 1876, she caught fire and ran aground again, just out of Guaymas. This time she was a total loss. Undaunted, the Colorado Steam Navigation Company promptly purchased her sister ship, the 1,077-ton *Idaho,* putting her on the run in January 1877. The *Idaho* was commanded by Captain George M. Douglass, who had replaced McDonough on the *Montana* after her first accident.[45]

Even with the risk of shipwreck the sea route to Arizona was still quicker, cheaper and far more attractive than the overland trek across the Mojave or Colorado deserts. Passenger fare from San Francisco to Yuma was $90 for first cabin and $40 for steerage. In the boom years of the 1870s the Colorado steamers were carrying more than a hundred passengers a month, and the accommodations were good enough so that a few adventuresome souls even took the trip purely for pleasure.[46]

Passage:

YUMA
—TO—
SAN FRANCISCO,
......PER......

Colorado S. N. Co's Steamers:

CABIN, $90 COIN   STEERAGE $40 COIN

FREIGHT ON WOOL,

YUMA TO SAN FRANCISCO

$40.00 PER TON.

Accommodtions, FIRST CLASS

I. POLHAMUS, Jr.,
feb26-tf.   Gen. Supt.

In the early 1870s the Colorado Steam Navigation Company bought two oceangoing steamers, the *Newbern* and the *Montana,* from the Pacific Mail Steamship Company, to run between San Francisco and the mouth of the Colorado. The ill-fated *Montana* caught fire and ran aground just out of Guaymas in December 1876. Lithograph by Endicott & Co., of New York.

Notice to Shippers!

THE COLORADO STEAM NAVIGATION CO'S STEAMERS

Newbern and Montana

Leave San Francisco for Mexican Ports and Mouth of Colorado River,

EVERY TWENTY DAYS

Connecting with River Steamers

For points along the river

Freight Delivered at Yuma in twelve days from San Francisco.

Superior Pasenger Accommodations

Agencies of the Company at

10 Market street, San Francisco, Cal
Yuma and Ehrenberg, Arizona.

I. POLHAMUS, Jr.,
Gen. Supt.

At the mouth of the river the ocean steamers anchored just off the entrance to Isabel Slough. There the passengers and freight were transferred to the river steamers and barges. This was usually accomplished in half a day. Then as soon as the tide turned the steamers headed upriver with one or two barges in tow. The desolate tidal flats around the estuary, enlivened only with scattered patches of salt grass, left such dreary impressions upon the minds of many passengers, that the clumps of scraggly mesquite and spindly cottonwoods that they encountered above tidewater seemed a lush paradise in comparison. By the time the steamer was thirty miles upriver, the channel had narrowed from three miles in width at its mouth to barely half a mile. From here on the sandbars were so numerous that the steamers had to tie up every evening because navigating at night was too risky.[47]

The steamers stopped dutifully for wood at Port Famine and the Gridiron, but aside from an occasional Cocopah rancheria there was little sign of life on the lower river until the steamer reached Lerdo Landing eighty miles above the mouth. Here on a mesa just east of the river sat Colonia Lerdo, a pleasant little town with eucalyptus-lined streets and some seventy inhabitants. The colony was started in June 1873 by San Francisco capitalist Thomas H. Blythe and Mexican Gen. Guillermo Andrade, who dreamed of building an agricultural empire on the lower Colorado, harvesting the wild hemp, or marijuana, that abounded in the bottomlands. The colonists also experimented with growing cotton and raised small patches of corn and vegetables for their own use.[48]

For many miles above Lerdo the only stops were again at the woodyards: Ogden's Landing, Hualapai Smith's and Pedrick's. To the west, the New River, a distributary of the Colorado, branched off toward Volcano Lake. Some forty miles down this channel was Poole's Landing, back of which Captain William Poole had a sulfur mine in the Cocopah Mountains from which he shipped ore every now and then. Just below the U. S. boundary the steamer passed another of General Andrade's enterprises, the Algodones colony, a vast tract of rich bottomland on which he hoped to settle some five hundred families from Sonora. Here the steamer followed the river's sharp bend to the east and in just a few miles reached Yuma—"the Seaport of Arizona."[49]

When the steamer arrived at the landing the seemingly sleepy town came to life instantly. The streets were suddenly jammed with "desert schooners," lined up to load freight for the interior. Indian and Mexican porters rushed back and forth carrying freight from the barges to the wagons. Passengers hurried ashore to be greeted by anxious friends. Children, loafers and street arabs crowded around to watch the excitement, and through the crowd wended the peanut peddler, the apple boy and the tamale vendor. On a busy day two or three steamers and as many barges would all be tied up at the landing unloading and loading freight.[50]

Arizona City, renamed Yuma in February 1873, had become the largest town on the river and the third largest in the territory. It boasted a population of some twelve hundred—two-thirds of whom were Sonorans. About five hundred Yumas lived on the

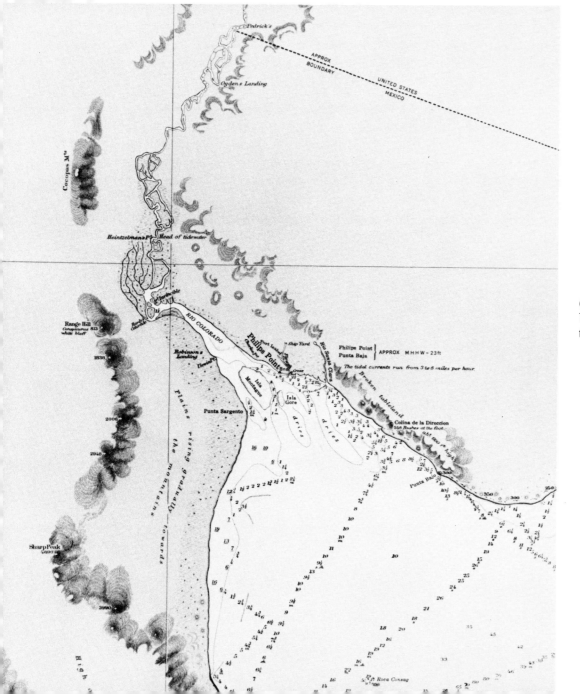

Chart of the mouth of the Colorado, survey by the U.S.S. *Narragansett* in 1873–75, showing the location of Isabel Slough and the Colorado Steam Navigation Company's shipyard at Port Isabel.

outskirts. It was the principal supply point for southern Arizona. Stages and freight wagons left almost daily for Phoenix, Tucson and other settlements in the interior. Yuma had also become the county seat in 1870 after the decline of La Paz, and in 1876 the Arizona Territorial Prison was established there.[51]

Despite Yuma's prominence, however, its low sunburnt adobes and dusty treeless streets gave the town a deceptively sleepy, lethargic appearance. In addition to the usual assortment of merchants and tradesmen, Yuma boasted those two vital adjuncts of civilization, a newspaper, the *Arizona Sentinel,* and a brewery, and as both a river town and an army town it naturally attracted more than the usual number of roughs, gamblers and prostitutes.

Steamers left Yuma every few days for the upriver landings, usually booked up with passengers and towing a bargeload of freight. Cabin fare upriver was five dollars from Yuma to Castle Dome, fifteen to Ehrenberg, twenty-eight to Aubrey, thirty-five to Fort Mohave and Hardyville, and forty-five dollars to Eldorado Canyon. The deck fare was about two thirds of the cabin fare. Both cabin and deck passengers were given meals and a mattress and straw pillow for bedding. The spartan cabins offered privacy and shelter from occasional rain but little else. Located on the upper deck above the boiler and engine, they were always stuffy and noisy, and in the summer they were so suffocatingly hot that their occupants abandoned them at night to spread their mattresses on the cool zinc deck with their second-class companions. The cabin passengers, however, did have the honor of eating at the captain's table, but the food generally was as spartan as the accommodations. Though they were occasionally treated to fresh meat and vegetables, dinner most often consisted of butterless biscuits, salt-boiled beef or pork and "Arizona strawberries" —dry beans. The one persistent bright spot on the menu of the *Gila* was a tasty peach and plum pie concocted by William Sam, the Chinese cook. The river water used for drinking had to stand for several days to settle out the fine silt, after which it was cooled somewhat in porous canteens hung from the deck railing. The traveler who wanted anything else to drink had to bring it with him.[52]

Scattered ranches and Yuma rancherias dotted the rich bottomlands on both sides of the river above Yuma. The first landing was Laguna on the Arizona side, 20 miles up. Here José Redondo had a store and ranch to which he had returned after the La Paz rush. Two miles below on the California side were the Pot Holes placers, which were only worked intermittently. The steamers made their first regular stop 35 miles above Yuma at Castle Dome Landing, supply center for the mining district of the same name. It contained a single store, a half dozen adobes and the furnaces of the Castle Dome Smelting Company which reduced the lead ore from their mines a couple dozen miles east of the river. Captain Polhamus also dabbled in mining in the district as half owner of the Flora Temple mine which shipped out hundreds of tons of galena ore by steamer to the smelter in San Francisco.[53]

Above here the river narrowed, coming in from the west through the Chocolate Mountains. A dozen miles farther, on the

Barges, such as the *No. 2* shown here with the *Mohave* (I), were also used as temporary wharves at the landings.

Arizona side, the steamer passed Eureka Landing and Williamsport—two small clusters of adobes a couple of miles apart—which supplied the mines in the Eureka district. On the opposite side of the river was Picacho, or Chimney Peak, Landing, with several adobes and huts nestled in a grove of cottonwoods and willows. Here David Neahr, the chief engineer for the steamboat company, had erected a five stamp mill to crush gold ore from the Picacho mine six miles back from the river. Built of quarried limestone blocks Neahr's mill was touted as "the handsomest structure on the river."[54]

Passing through Red Rock Gate the mountains recede from the river, and a valley opens out stretching for more than a hundred miles to the north. On the Arizona side bordering the river for sixteen miles was Rood's Ranch, the largest and oldest on the upper river. William B. Rood, one of the Death Valley forty-niners, brought a herd of cattle here in 1862 to supply beef to the La Paz diggings. He was drowned in April 1870, while crossing the river, and his ranch passed into the hands of José Redondo and his brothers.[55]

Navigation became difficult here as the river fanned out over a myriad of sandbars, reaching a width of a third of a mile. At low water it was only a foot or so deep in most places. Finding a channel deep enough for a steamboat in this labyrinth of bars was often a frustrating task. The captain was constantly at the wheel. At the bow the leadsman took soundings with a long pole and sang out the depth. When it exceeded six feet he was silent, but all too frequently he would sing, "Four! Three! Two! Two light! Quarter less two!" and then with a jolt the steamer was on a bar. The deck hands would swing over the side into the river with poles to help push the boat off, as the captain rang the engineer to reverse the engine. If he was unable to spot a deeper channel through the muddy water, the captain would take the dinghy ahead to try to sound out one with a pole. If this failed he simply had to work the boat over the bar one way or another.[56]

First, he usually tried to "grasshopper" the boat over with poles and spars. Running full ahead onto the bar as far as he could, he called out, "Get out your small boat, take the big anchor and four-inch line to the head of the bar!" While this was being done he hollered to the deck hands, "Set your spar on starboard side." They drove the spar deep into the bar, then ran a line from the other end through a pulley on the side of the boat to the capstan. With the click! click! click! of the capstan poles, the starboard side of the steamer was slowly raised. The captain rang the engineer to "Go ahead." The stern wheel churned up a froth, the men at the capstan took up the slack on the anchor line, and the boat slipped forward a few feet. Then the spar was reset and the process repeated again and again till the boat was over the bar.[57]

If the water over the bar was too shallow, however, the captain had to bring the boat around stern to and "crawfish" over, cutting a channel with the stern wheel. The novelty of this technique, devised by Captain Mellon, impressed the passengers no end. "It is pure Yankee!" one Swiss traveler exclaimed. To turn the boat stern to, the anchor was set at the head of the bar and the line run through a pulley at the stern then forward to the

John Alexander Mellon, the best known of the Colorado steamboat captains, came to the river with Trueworthy in 1864 and stayed for nearly half a century.

capstan. With the help of the current, the boat swung around end for end as the capstan turned. Then the captain rang the engineer to full reverse the engine, and taking up the slack on the anchor line as he went, he dug and pulled his way through the bar with the paddle wheel. In this way the boat could be taken over a bar on which there was as little as two inches of water! Sometimes, of course, there was not even this much, and in trying to get over the steamer would run solidly aground, there to remain until a temporary rise in the river. The steamboats were rarely aground more than a week at a time, though on one occasion Jack Mellon was caught for fifty-two days![58]

Such incidents provided the only excitement to relieve the monotony of the trip, but even running aground soon became much more aggravating than exciting as it happened again and again. This part of the river was also plagued with sandstorms from the open desert to the west, choking the air and obscuring the sun for days at a time. A feeling of oppression and despair often gripped the passengers unable to escape the fine grit and sand which found its way into every crevice of the boat, every piece of bedding, every bite of food and every breath of air. The accompanying wind was also so strong that the captain, unable to hold the steamer in the channel, had to tie up to the bank until the storm abated.[59]

For nearly sixty miles above Rood's Ranch there were only occasional woodyards and a couple of ranches with landings called California Camp and Drift Desert. Finally, 125 miles and three or more days above Yuma the steamer reached Ehrenberg, the second largest town on the river and the shipping point for central Arizona. Most of the upriver freight was landed here, and most of the passengers disembarked to take stagecoaches to the interior. As at Yuma the arrival of the steamer at the landing brought forth a rush of activity as the desert schooners loaded up with freight for Prescott, Wickenburg and the surrounding mines. The town, made up entirely of low dusty adobes, contained from four to five hundred inhabitants, most of whom had moved there from La Paz after the decline of the placers in the late 1860s. To an eastern lady who had to live there, Ehrenberg seemed "an unfriendly, dirty, heaven-forsaken place," but to the thirsty steamboat passenger it offered the first chance in days to drink something other than tepid river water.[60]

The Colorado River Indian Reservation covers both sides of the river for eighty-five miles above Ehrenberg. It was laid out in 1864 by Charles Poston, for the Mohaves, Chemehuevis, Hualapais and Yavapais, but since the latter two tribes were hunters, not farmers, they refused to settle there. Johnny Moss, trying to get a hand into the pork barrel, took Mohave Chief Iretaba on a tour of the eastern states in an unsuccessful bid to be named agent for the river Indians. The Indian agent lived at Parker's Landing, four miles below the present town of that name, and to impress the Indians with the power of the government a detachment of troops was stationed at nearby Camp Colorado until its brush huts were burned out in 1869 by a stray spark from the steamer *Cocopah*.[61]

The next landing was at Empire Flat on the Arizona side, ten miles above Parker's. It consisted of a little five-stamp mill, a mess hall and a few huts for the mill hands. The mill was a two-story open structure, built to work gold-bearing copper ore from the mines in the Buckskin Hills just back of the river. The superintendent slept on the top floor hoping to keep away from rattlesnakes and scorpions.[62]

Just around the bend above, the steamer came to the mouth of Bill Williams' Fork. Though it was dry most of the year, it was, nonetheless, the only significant tributary of the Colorado between the Gila and Virgin rivers. At the junction was Aubrey Landing, named for the so-called "Skimmer of the Plains," Francois Xavier Aubrey, who had forded the river at that point in 1853. At the peak of the river rush in the mid 1860s, Aubrey "City" had boasted some fifty cabins and stores, but it was virtually abandoned with the first collapse of copper prices in 1865. A decade later only four or five cabins remained, the largest of which served as a combination post office, store, saloon and hotel. An old ship's cabin served as the office of the Colorado Steam Navigation Company and the other huts saw life only when the freight teams were in, loading supplies for the mines and camps on up Bill Williams' Fork. The Planet mine, a dozen miles up the fork, shipped copper ore on nearly every steamer that had gone downriver since 1864. In the hills north of the fork was Jackson McCrackin's gigantic silver lode, discovered in September 1873. McCrackin sold the mine to Nevada senators J. P.

The *Gila* stopping at Castle Dome Landing on her way upriver in the 1870s.

Jones and W. M. Stewart for $24,000. They moved the old Moss mill over to Big Sandy Creek, just east of the mine, in 1875 and ultimately took more than $6 million out of it. Greenwood City sprang up around the mill, and within two years two more mills were built on down the creek and the rival camps of Signal and Virginia were founded. They had a combined population of nearly a thousand for a time, all supplied by steamer and wagon from the landing at Aubrey.[63]

The river above Aubrey passed through Chemehuevis Valley, where Captain Isaac Polhamus and a few others had started ranches. Chimehuevis and Liverpool landings, serving both the ranchers and scattered miners, were the only regular stops for the steamers until they reached Fort Mohave.[64]

Finally some 310 miles above Yuma the steamers tied up at Hardyville, which was for most of the year the practical head of steam navigation. Though it never boasted more than a hundred residents, Hardyville was, nonetheless, the principal shipping center for northern Arizona and for several years the Mohave County seat. The mines in the mountains to the east had been abandoned during the Hualapai War which had erupted after the treacherous murder of Chief Wauba Yuma in April 1866. Late in 1870, however, as the trouble subsided, a band of "fighting miners" moved back into the Cerbat Range to reopen the mines. Within a year more than five hundred miners had flocked into the district. Even steamer Captain A. D. Johnston and bargeman L. C. Wilburn were drawn to the mines. Three camps, Cerbat, Mineral Park and Chloride, vied for the title of "boss camp."

Cerbat won the first round, taking the county seat away from Hardyville in 1873, but Mineral Park grabbed the honor four years later. The Hualapai mines were the mainstay of Mohave County for the next two decades. Though a couple of mills were erected to work the ore locally, their recovery was so poor that much of the high-grade ore was shipped out to San Francisco at a cost of forty dollars a ton—twenty dollars a ton by wagon 35 miles to Hardyville and twenty dollars by steamer the remaining 2000 miles.[65]

Only a few steamers a year ventured above Hardyville to carry wood and supplies up to Eldorado Canyon during the summer high water. The rest of the time the camp was supplied mainly from the Mormon settlements upriver by flatboats which were broken up and sold as firewood on arrival. The mines in the canyon underwent several changes of management, but they were worked steadily.[66]

After the abandonment of Callville, Eldorado Canyon was the only important settlement on the upper river until the turn of the century—though there were two paper towns which flourished briefly in the imaginations of their creators. The first was Freemansburg, the brainchild of a peripatetic newspaperman, Legh R. Freeman, who platted it at the mouth of the Virgin River in February 1868, proclaiming it to be not only the head of steam navigation but the "Sanctum of the American Libertarian" and the "Capital of the Territory of Aztec!" Two years later the ubiquitous Johnny Moss envisioned a second paper town, Piute City, on the west bank of the river opposite Cottonwood Island. Its

The *Gila* brought bargeloads of coal up to the stamp mill at the mouth of Eldorado Canyon several times a year.

Piute City, a figment of the artist's imagination, was conjured up for the prospectus of one of Johnny Moss's mining company promotions. From a lithograph by George H. Baker.

existence never went beyond an attractive artist's rendering in the prospectus for The Piute Company of California and Nevada, which showed it as the bustling shipping point for the company's mines at Ivanpah, forty miles to the west. Though these mines were later developed, their supplies were carried overland from San Bernardino.[67]

By the mid 1870s the Colorado Steam Navigation Company was doing over a quarter of a million dollars a year worth of business, handling some seven thousand tons of freight and about a thousand passengers annually. More than four-fifths of this income came from inbound freight—the tools, explosives, and machinery for the mines, the stamps, boilers and heavy timbers for the mills, the supplies and equipment for the army, and the dry goods, hardware, furniture, and even food for much of the territory. The bulk of the freight was carried from San Francisco to Yuma at fifty dollars a ton and on upriver at additional charges of as much as sixty dollars a ton to Eldorado Canyon. The freight charges on heavy goods were based on actual weight, whereas those on lighter freight were based on volume, with forty cubic feet being defined as a "measured ton." A piece of furniture in a four-foot-square crate would thus cost the same to ship as 3200 pounds of mining machinery. This practice obviously worked to the advantage of the mines but was the cause of continual complaint from Arizona merchants. The outbound freight, which provided only a small percentage of the company's income, was made up primarily of silver, lead and copper ore with seasonal shipments of wool, hides and pelts. This was carried effectively as ballast on the ocean steamers, and the downriver freighting rates were much less than those upriver, further helping to stimulate mining along the river. Miners could thus ship their ore to San Francisco from Yuma for only $10.00 a ton—only a fifth of the upriver rate—from Ehrenberg for $12.50 a ton, from Bill Williams' Fork for $15 and from Hardyville for $20. The remainder of the company's earnings came from fares of passengers, more than half of whom were soldiers going to and from posts in Arizona.[68]

Indispensable as the river steamers were to the Arizona trade, more than half of the Colorado Steam Navigation Company's earnings came, not from the steamers working the river, but from the company's seagoing steamers on the run from San Francisco to the mouth of the river. Thus when the railroad finally reached Arizona in the late 1870s the steamboat company immediately lost over half its income and the bonanza days of the Arizona Fleet ended. Steamboating on the river, however, continued till well after the turn of the century.

On 30 September 1877 the first locomotive rolled across the Southern Pacific railroad bridge into Yuma, breaking the steamboat monopoly on Arizona trade. For the next three decades the steamers played an ever diminishing role as the railroads slowly usurped their trade.

Through Progress of the Railroads

"Through progress of the railroads, our occupation's gone..." was the ubiquitous lament of the western teamster. It was no less common with the western steamboatman, for in the long run neither could compete with the iron horse. The progress of the railroads was first felt on the Colorado River in the spring of 1877 with the extension of the Southern Pacific railroad to Yuma, which broke the Colorado Steam Navigation Company's monopoly of Arizona trade. Coastwise, shipping to the mouth of the river came to an abrupt end with the coming of the railroad, and river trade above Yuma was soon reduced to only a fraction of what it had been. Sustained, however, by new mining developments along the river and the introduction of two new types of boats—the gasoline-powered launch and the massive gold dredge—the river business continued for another quarter of a century. As the railroads pushed farther and farther into the upriver country, however, they slowly eroded the river trade until, at last, the damming of the Colorado cut it off entirely.

The Southern Pacific tracks were completed across the California desert to Jaeger's ferry on the banks of the Colorado River in May 1877. The railroad company offered to carry freight from San Francisco to Arizona in only three days for forty-two dollars a ton, and shippers lost no time taking advantage of the faster, cheaper rates. To handle the enormous volume of freight, Colorado Station, a temporary transfer point, was established at the ferry, while a bridge was being built a mile upriver. The Colorado Steam Navigation Company's *Barge No. 2* was tied up at the station as a wharf, and the old steamer *Colorado* was put back in service as the "transfer boat" to ferry freight and passengers across the river to Yuma. In the meantime the *Cocopah, Gila* and *Mohave* were all kept busy carrying the railroad freight on upriver from Yuma. In defiance of an order by the secretary of war the tracks were laid across the bridge at Fort Yuma and in a flurry of excitement early Sunday morning 30 September 1877 the first locomotive rolled across the railroad bridge over the

Colorado into Arizona—opening a new era in the history of the territory and dooming another on the river.[1]

Realizing the futility of fighting the railroad, George Johnson and his partners had sold out their interest in the Colorado Steam Navigation Company to the Western Development Company, a holding company for the Southern Pacific's backers—Stanford, Huntington, Hopkins, Crocker and Colton—on 21 May 1877 just as the railroad reached the river. Johnson retired from the shipping business completely to devote his full energies to his ranch and racehorses. Two of his partners, Alfred Wilcox and John Bermingham, who had previously bought part of Johnson's interest, kept the ocean steamer, *Newbern,* and formed the California & Mexican Steamship Line to continue the still profitable portion of the coastwise trade from San Francisco to the Mexican ports of La Paz, Mazatlan and Guaymas.[2]

The new owners of the Colorado Steam Navigation Company carried on the river business as one of their many private adjuncts to the railroad business. They kept Superintendent Isaac Polhamus and veteran captains Jack Mellon, Charles Overman and Steve Thorne, but they made some controversial changes in the operation of the company. Attempting to reduce costs, they discharged all the deckhands working by the month and rehired them only by the trip at a miserly eighty-three cents a day. This, in effect, cut their wages by more than half and most refused to go back to work for the "Godless outfit." As a result steamer departures were postponed for several days for lack of crews, and the company experimented unsuccessfully with Chinese crews before they were finally forced to resume the old system. In another unpopular move the company instituted a $5.00 fare for dogs, which had previously been carried free with their masters. They were forced, however, to compromise on this issue, too. In the face of protest by dog lovers, they reduced the charge to $2.50. This won them tongue-in-cheek praise from the local press for showing "such liberality . . . in meeting our people half way in developing the resources of Arizona."[3]

The most significant change made by the new owners, however, was the abandonment of Port Isabel—no longer needed since the coastwise steamers had stopped calling at the mouth of the river after the advent of the railroad. The shipyard was dismantled and transferred to Yuma early in 1878. The company constructed new ways for the boats, as well as blacksmith's and carpenter's shops with steam-powered bellows and lathes. The ways at Yuma could not be used during high water, however, so Captain Polhamus took one of the boats downriver two years later to overhaul her in the old dry dock at Port Isabel, but he found it full of sand. Only the broken-down hulks of a couple abandoned steamers remained as a ghostly reminder of the once busy port. Pilings, scattered timbers and scrap iron mired in the mud still mark the spot.[4]

The abandonment of steamer service below Yuma worked a hardship on the struggling Lerdo colony and led its backers, Thomas Blythe and Guillermo Andrade, to form the Gulf of California Steamship Company to reestablish service. They purchased two boats in May 1878 to run all the way from Lerdo down

Isaac Polhamus, Jr., shown seated on the right outside the steamboat office in the late 1870s, continued as superintendent under the railroad management until 1886 when he and Jack Mellon bought the boats.

the gulf to San Blas, and they contracted with the Mexican government to carry the mail between intermediate ports. In July the seagoing steamer *General Zaragosa* began a semi-monthly run from San Blas to the mouth of the river, touching at Mazatlan, Altar, La Paz, Mulege, Guaymas, Port Libertad and San Felipe. Built in 1851 as a side-wheeler, she had run for many years on the Sacramento and San Joaquin rivers as the *C. M. Weber* and on San Francisco Bay as the *Guadalupe*. Afterward she was converted to a seagoing propeller as the *Coquille,* then sold to Blythe for $30,000 and renamed the *General Zaragosa* for the Mexican general who defeated the French at Puebla in 1862.[5]

At the mouth of the river the *General Zaragosa* connected with a small propeller-driven steam launch, the *General Rosales,* commanded by Captain C. A. Eastman, which ran up the river to Lerdo. The *Zaragosa* made only a few trips up the gulf, however, before the Mexican government cancelled the mail contract in August after learning that she was less than a quarter of the tonnage called for in the contract. Blythe and Andrade needed the mail to sustain operating costs, so the cancellation of the contract killed the business on both the gulf and the river. The *General Rosales* was taken to Guaymas and the Lerdo colony was left to rely thereafter on wagons from Yuma and an occasional sailing sloop venturing up from the gulf ports. Blythe and Andrade talked briefly of building a Yuma & Port Isabel Railroad through Lerdo, but it came to naught.[6]

Throughout 1877 and 1878 the Colorado Steam Navigation Company did a prosperous business carrying nearly all the freight for northern and central Arizona from the railroad at Yuma to the transshipping points upriver. This amounted to roughly ten thousand tons of freight a year and paid the new owners a couple of handsome dividends. But the construction crews slowly pushed the railhead east, so that by April 1879 trains were running halfway across Arizona. Maricopa Wells, 190 miles east of Yuma, then became the new transshipping point for central Arizona, and the river business sharply declined. Where the previous year four boats were hard pressed to handle the freight, now there was not enough to keep one boat busy. The *Colorado* and *Cocopah* were retired and two years later dismantled, leaving only the *Gila* and *Mohave* (II) to work the river with a couple barges.[7]

The river towns were hard hit by the loss of trade. The population of Yuma fell from around fifteen hundred to only about five hundred; every line of business was stagnated; and the local press damned the Southern Pacific owners for having "practically withdrawn their steamers from the river." At Ehrenberg the last inhabitants were ripping the windows and doors from their adobes and heading for the interior. Once the most populous point upriver, it was virtually abandoned, the refuge only for a few derelicts. Johnson's former partners congratulated themselves for their foresight, as John Bermingham wrote to Captain Wilcox in May 1879, "The river business is used up. . . . We got out of the business just in time."[8]

Ironically it was at this same time that the Colorado steamboats finally reached that long sought "head of navigation"—the

The Southwestern Mining Company's stamp mill at Eldorado Canyon was the principal mining operation on the river during the 1880s and early 1890s.

mouth of the Virgin River. The biggest mining operation on the Colorado River in 1879 was the newly organized Southwestern Mining Company. Headed by mining entrepreneur Joseph Wharton, of Philadelphia, the company consolidated most of the paying mines in Eldorado Canyon, completely overhauled the old mill, and chartered the *Gila* for four months to bring in new machinery. Jack Mellon captained the boat while under charter, and as soon as he had delivered the first cargo of machinery, he was ordered to try to take the *Gila* on upriver to the salt mines and the Mormon settlement of Rioville at the mouth of the Virgin. He set out from Eldorado Canyon at 8:30 A.M. on 7 July, and making record time through the rapids in Black Canyon, he tied up at Callville that evening. The following day he carefully guided the *Gila* the last twenty miles through the uncharted rapids of Boulder Canyon to his destination.[9]

On 8 July 1879, twenty years after Johnson and Ives first set out for the elusive Virgin, Jack Mellon finally proved that it was indeed the head of steam navigation on the Colorado River. The Mormons at Rioville were "wonder-struck" to see a steamboat; one proclaimed it the "biggest thing he ever saw in water." The passengers on the *Gila* were equally impressed with the deep canyons through which they had come, vowing they were the "grandest on Earth." They felt dwarfed by the towering walls, and one concluded that the steamer could not have looked more out of place in the bottom of a mine shaft.[10]

During the next eight years Mellon took the *Gila* back up to the Virgin a couple times a year to get salt for the mill, eventually making a total of twenty trips. Captain Isaac Polhamus also took her up a couple times, and his nephew Captain Joseph H. Godfrey even succeeded in working the *Mohave* up to the Virgin River once in the summer of 1881. To help the boats through the worst rapids, Mellon secured half a dozen ringbolts to the canyon walls in 1883. However, since the steamers could only make the trip during high water and, in fact, at low water could not even reach Eldorado Canyon, the mining company bought a sloop, the *Sou'Wester*, to make the run the rest of the year. She was a fine, swift boat of 65 tons burden, carrying 18 tons of salt on 2 feet of water. She had a cedar deck and a boiler-plated hull, 56-foot keel and 15-foot beam, and a 48-foot mast sporting more than 400 square yards of canvas. Built in San Francisco, she was shipped by rail to Yuma, then upriver on the *Gila* to Mormon Island, where she was reassembled and launched in November 1879. With Captain Mellon at the helm, the *Sou'Wester* made nineteen trips to the Virgin before she was wrecked by his first mate in the Short and Dirty Rapids in 1882. Steamboating above Eldorado Canyon finally ceased in 1887 with the decline of mining operations.[11]

The only other important mining activity upriver was the continuing work around Mineral Park and Signal. To shorten the overland haul to Mineral Park and break Bill Hardy's monopoly on the trade, two of the camp's merchants in partnership with the steamboat superintendent established Polhamus Landing some five miles above Hardyville in May 1881. Within a month a large warehouse was built, and thereafter more freight was landed

The Eldorado Canyon trade even attracted the "palatial" *Mohave* (II) as this half-page advertisement in the *Arizona Sentinel* of 1881 shows.

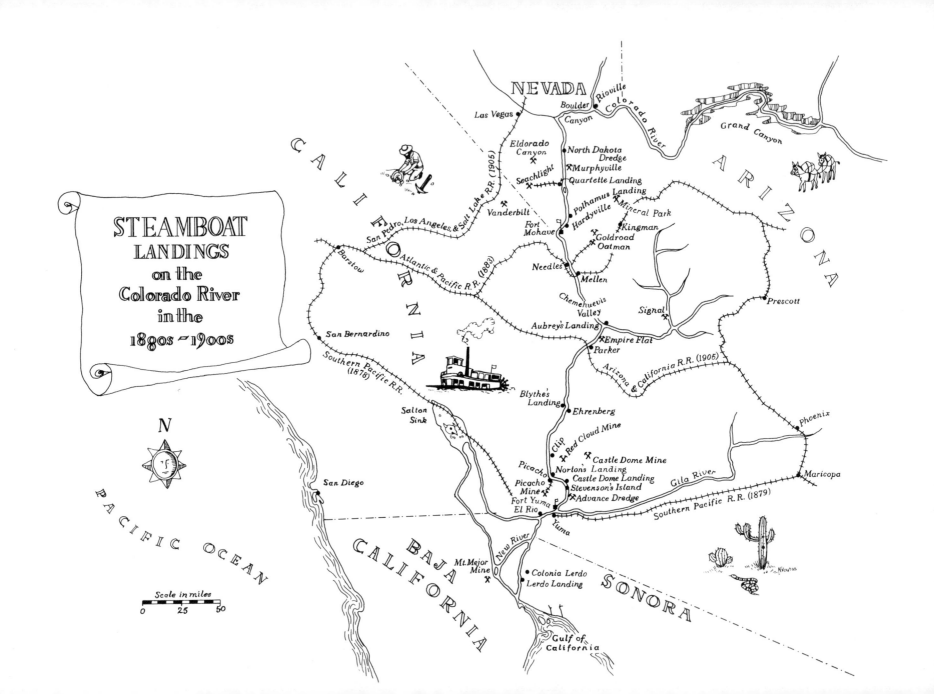

there than at any other point on the river. On down the river some freight was still landed at Fort Mohave and at Aubrey for the mines around Signal, but only a wood stop was usually made at the once busy landing at Ehrenberg.[12]

On the California side of the river, opposite Ehrenberg, however, there was brief activity. There Thomas Blythe, promoter of the Lerdo Colony in Mexico, had started a second colony in 1879, the first such settlement on the river north of the border. He claimed a total of 80,000 acres of rich bottomlands as Swamp Land District No. 310, paying the state one dollar an acre. Here he envisioned a new Dixie of big plantations growing sugarcane, cotton and tobacco, irrigated by a grand canal, more than 120 feet wide, and worked by Mohaves from the neighboring reservation. But the colony was plagued with trouble. The Chemehuevis still claimed the land as their planting grounds and in March 1880 Blythe's manager, O. P. Calloway, and one of the Chemehuevis killed one another over the dispute. Little progress had been made three years later when Blythe died suddenly of a "fainting fit." The lengthy litigation over his estate further impeded development of the property for several years after that. The river landing, first dubbed 310 Landing, was renamed for Blythe after his death. The name later passed to the present town when the bottomlands were finally developed along less feudal lines.[13]

The mines in the Cargo Muchacho, Picacho, Silver and Castle Dome districts, just above Yuma, were the only other producing mines on the river at that time, and only the Castle Dome mines provided any significant business for the steamers.

Starting in the late 1870s the Castle Dome mines shipped to their smelter near Oakland nearly one hundred thousand tons of lead ore on which the steamboat company collected $1.50 a ton for the thirty-five mile run to Yuma. The Picacho gold mines on the other hand could work their ore at their mill on the river, so neither their supplies nor bullion shipments netted the steamboat company much revenue. Similarly the revival of mining in the adjacent Silver district provided little steady income for the steamers since these ores, too, were worked locally. First located by old Jacob Snively in the mid 1860s, these lodes attracted little attention until George W. Norton relocated them in 1878. Within two years the Red Cloud Mining Company had built a successful reduction works at Norton's Landing, known briefly as Pacific City, and the Clip Mining Company built a ten-stamp mill eighteen miles upriver a couple years later. The Cargo Muchacho mine located in the mountains of the same name, twenty miles northwest of Yuma, had been worked only intermittently after their discovery in 1862, until W. W. Van Arsdale got control of the property in the late 1870s. With British capital he put up a ten-stamp mill at El Rio on the Colorado five miles west of Yuma. The first stamps dropped on 30 June 1879 and the mine began shipping thirty to forty tons of ore a day, but the river steamers got none of this business because the Southern Pacific Railroad passed within three miles of the mine and a spur track was run right to the mill.[14]

The coming of the railroad had brought reduced shipping cost throughout most of the Arizona Territory, providing a great stimulus to mining there, but it was of no benefit to the mines adjacent to the Colorado where shipping rates remained the same as those charged by the old steamboat company. Thus as the rest of the territory boomed in the late 1870s and early 1880s with the expansion of the railroad network into the interior, mining along the river languished. Only the upriver trade to the mines at Eldorado Canyon and Mineral Park and the army at Fort Mohave sustained the riverboats and even much of this was soon lost when the Atlantic and Pacific Railroad, now the Santa Fe, crossed the upper river.[15]

In May 1883 construction crews, laying the Atlantic and Pacific tracks west across Arizona and east from California, met at the Colorado River twenty-five miles below Fort Mohave. The tent town on the California side, dubbed Needles for the pinnacles downriver, quickly became the largest port on the river above Yuma. Here the railroad company built a hotel, car sheds, shops and a roundhouse. Within a month the town also boasted a Chinese washhouse, a newsstand, a restaurant, a couple of general stores, and nine or ten saloons, dispensing whiskey at two bits a drink.[16]

For three months after they reached the river the crews struggled to bridge the Colorado just below Needles. It was an ill-suited time and place for building a bridge. The river was at flood stage and the channel was 1600 feet wide at that point with no solid banks on either side. The swift current uprooted the pilings almost as fast as they could be driven. A wide gap mid-stream resisted all efforts before it was finally conquered with the aid of the *Mohave* and a pile driver mounted on *Barge No. 3*. Even then the bridge was criticized as a "flimsy looking structure," and an obstruction to navigation since it lacked a draw. The precariousness of the site was demonstrated again and again as the Colorado floods swept away the bridge in 1884 and its two successors in 1886 and 1888. The Atlantic and Pacific company belatedly recognized their error after the loss of their third bridge and began construction of a high cantilever span at a much narrower point ten miles downstream. This bridge was completed in May 1890, and the station there was named Mellen in misspelled honor of the steamer captain.[17]

Before even the first bridge was completed, however, Captain Mellon had taken the *Mohave* and her barge above the bridge to carry freight on upriver above Needles while the *Gila* under Captain Polhamus remained below the bridge to handle the trade on that stretch. But the new railroad, which passed close to the mines around Mineral Park, cut heavily into the remaining steamboat trade upriver. There was, in fact, so little river trade that after a draw was put in the bridge in 1884, the *Mohave* was taken back down to Yuma and tied up. The *Gila* was the only boat in service then, and she was making trips upriver only once every six weeks or so. Reflecting on the fact that before the "advent of the railroad" the Colorado Steam Navigation Company had operated as many as five steamers and five barges at one time, the Yuma *Sentinel* editor lamented, "How the mighty have fallen. From a powerful corporation it has been reduced almost to naught. . . . Water transportation can never compete with railroads." Even the Southern Pacific's backers were no longer

Construction of the Atlantic and Pacific railroad's cantilever bridge across the Colorado below Needles gave the *Gila* only temporary work in 1889 as the railroad itself cut further into the declining river trade.

interested in the riverboats and on 10 September 1886 they sold out their interest in the failing company to Isaac Polhamus and Jack Mellon.[18]

For the next half dozen years Polhamus and Mellon barely kept the business afloat. Carrying both supplies to the Indian agencies at Parker and Fort Mohave, and coal for the stamp mills at Eldorado Canyon and Hardyville made up the bulk of the business for the steamers. Added to this were only occasional supply and machinery shipments to the other mines at Castle Dome, Norton's Landing, Clip and the Black Metal Ledge near Aubrey, and the scattered ranches at Blythe's Landing and Chemehuevis Valley. May Day and Fourth of July excursions from Yuma to picnic grounds on Stevenson's Island some twenty-eight miles upriver and from Needles to Fort Mohave also helped supplement the steamer revenue. In an effort to attract transcontinental railroad travelers, Polhamus and Mellon tried unsuccessfully to promote regular steamboat tours down to the Gulf of California and up to the foot of the Grand Canyon, offering such enticements as, "Think of passing through canyons from the profound depths of which shining stars can be seen from the deck of the steamer at midday!"[19]

The decline of the river trade in the late 1880s resulted not only from ever-increasing railroad competition, but also from ever-declining silver prices which forced the closing of many mines and which culminated in the virtual collapse of silver mining throughout the West, following the repeal of the Sherman Silver Purchase Act in 1893. The collapse of silver, however, finally spurred prospectors back into the hills to look for gold, and rich new discoveries soon revitalized the mining industry.

Starting in the early 1890s a number of new mining districts were opened up along the Colorado River. In January 1891 a Piute, Robert Black, discovered gold in the New York Mountains, thirty-five miles from the river, northwest of Needles. A rush began, and the town of Vanderbilt sprang into existence. A few months later gold was found on the Arizona side of the river just twelve miles below Eldorado Canyon at what soon became known as Murphyville. The following year another boom camp was born, when rich ore was found at White Hills, some twenty miles east of Eldorado Canyon. At the same time gold ledges in the Picacho and Cargo Muchacho districts northwest of Yuma were reopened. These were followed during the next ten years by a flurry of other discoveries all along the river. The most productive were the mines at Searchlight, just fourteen miles west of the river, south of Eldorado Canyon, and the mines at Goldroad and Oatman, a dozen miles east of the river at Fort Mohave. The Searchlight mines, discovered in May 1897, yielded around $4 million in gold, and the Goldroad and Oatman mines discovered about the same time ultimately produced some $35 million.[20]

During the boom several new mills were built along the river and three short line railroads were run from different mines to the river mills. The first was the Quartette Mining Company railroad built in 1901–02 between Searchlight and the company's mill at Quartette on the west bank of the river about sixty miles above

The turn of the century mining boom started at Searchlight, which gave its name to the last steamboat on the lower Colorado, and brought a brief revival of the river trade.

Needles. At the same time the California King Gold Mines Company built a winding, five-mile railroad from their mines to a new cyaniding plant on the river at Picacho, and in 1903 the Mohave and Milltown Railway was run from the Leland and Vivian mines near Oatman to the Colorado opposite Needles. These railroads were not branches of the main lines, but were wholly dependent on the river, and everything from their ties and rails to their engines and coal were brought in by the steamers.[21]

The river trade prospered once again with the renewed mining activity. Polhamus and Mellon were once more able to keep both the *Gila* and *Mohave* busy, and soon they had to build a new barge, the *Enterprise,* to keep up with the business. The *Enterprise* was equipped with a small steam engine rigged to her capstan so that she could pull herself upriver.[22]

The increased business on the river, however, invited competition, and the growing popularity of the newly developed gasoline "vapor engine" led to the introduction of a number of smaller steam- and gasoline-powered boats. These rival boats, in fact, finally broke Polhamus and Mellon's monopoly of the river business. The first to take up the challenge were E. E. and O. T. Stacy of Yuma. In December 1891 the Stacy brothers launched the first gasoline-powered boat on the river, the *Electric Spark*. Before the end of the month she was carrying freight up to Castle Dome Landing and bringing down ore. She was an instant success, but it was soon apparent that she was over-powered for her size, so the Stacy brothers wasted no time building a larger boat for her engine.[23]

Their new boat, named simply the *Electric,* had a catamaran hull. She was launched early in March 1892, her engine was transferred, and within a week she was headed upriver on her maiden voyage to Ehrenberg and way landings with three passengers and a heavy cargo of supplies. With E. E. Stacy as her captain, the *Electric* was soon making almost weekly trips supplying the ranches and smaller mines up and down the river from Blythe Landing to the old Lerdo Colony. Her light draft also allowed her to reach Geronimo Elizaldo's new gold mine on Mt. Mejor in Baja California by running fifty-five miles up the New River from the Colorado. The *Electric* became a very popular boat, since Polhamus and Mellon's larger boats, which required bigger cargoes to be profitable, could not afford to make such frequent trips. Stacy also took excursion parties on short runs up the Gila, and in the fall of 1892 he chartered the *Electric* to Nevada bonanza king and former U.S. Senator James G. Fair for a three-week pleasure trip to the Colorado estuary.[24]

By the end of 1892 the Stacy brothers had more business than the little *Electric* could handle, even with a barge in tow, and thus they commenced building a bigger, more powerful boat, the *Aztec*. She was a 62-foot long stern-wheeler with a 21-foot beam, she was run by a 20 horsepower gasoline engine, and she had a capacity of 50 tons on 20 inches of water. The *Aztec* was launched at Yuma 13 February 1893 and made her maiden voyage to Castle Dome two months later. She joined the *Electric* making frequent runs to the upriver landings, moonlight cruises up the Gila and hunting and pleasure trips to the estuary in fall and

The Stacy brothers' gasoline-powered stern-wheeler *Aztec,* shown passing through the Southern Pacific railroad's swing bridge at Yuma in the early 1890s, gave Mellon and Polhamus further competition for light freight.

winter. On one occasion Stacy even braved the open waters of the Gulf of California in the *Aztec,* taking her over a hundred miles down the coast of Baja California on a five-week excursion to Bahia San Luis Gonzaga. The Stacys soon found, however, that they did not have enough business for both boats, so they retired the *Electric* late that year.[25]

Polhamus and Mellon, in the meantime, were feeling the pinch of the Stacy brothers' competition for the river freight and excursion business. Not to be outdone by an upstart, Captain Mellon even took the palatial *Mohave* out into the gulf on a ten-day excursion and would have made regular trips, but he could not sell enough tickets to pay expenses for the big boat. Finally in 1895 Polhamus and Mellon bought out the Stacy brothers. They put the *Aztec* to work on the run from Needles to Eldorado Canyon during low water, but she proved no more successful there than had their larger boats.[26]

Needles was prospering as the supply center for the new mines upriver and a group of eastern capitalists, led by Chicago railroad man Warren G. Purdy, proposed to make it a milling and smelting center as well. Early in 1892 they organized the Needles Reduction Company to erect a cyaniding works and smelting furnace just north of town. To bring the ore to their plants they also floated the Colorado River and Gulf Transportation Company, which commenced operations with a 25-foot steam launch on 30 June 1892. However, the editor of the *Needle Eye,* expecting something grander from the eastern entrepreneurs, complained that the boat was such a "small concern . . . we wonder how it will ever do any good on this peculiar stream!" His doubts proved well founded. Within a month the rival river operation was abandoned and Polhamus and Mellon were handling all the river freight for the reduction works.[27]

The Stacy brothers' success with gasoline boats at Yuma prompted similar experiments at Needles, but they all failed. The first attempt ended tragically. In May 1893 E. J. Oleson, who had put together a makeshift paddle-wheel boat with a small gasoline engine, took his family out in her, was hit and sunk by a drift log, and lost a child. A gasoline boat, built a few years later by Rice, Hill and Company to serve their new mill at Eldorado Canyon, was unable to stem the current and was abandoned at Needles. The *Little Dick,* built by E. S. Blaisdell of Empire Flat in the spring of 1897, made good time going downstream but broke her propeller going up.[28]

The little gasoline boats such as these had offered no serious competition for the bulk of the river trade, but beginning in 1899 Polhamus and Mellon faced more formidable challenges which ultimately broke their monopoly. The first came early in 1899 with the launching of a rival stern-wheel steamer, the *St. Vallier,* built at Needles for the Santa Ana Mining Company which had placer claims fifty miles upriver. The *St. Vallier* was a 74- by 17-foot, steel-hulled boat with a light draft and a powerful engine. She proved to be a fairly successful boat, but her early career was clouded by financial problems. The Santa Ana company overextended itself, and its creditors attached the boat soon after she was completed. They tied her up in litigation until the summer of

The *Aztec* also cut into the steamers' excursion business, but the musicians had to play as loud as they could to be heard above the gasoline engine.

The success of the *Aztec* led to similar experimentation with gasoline-powered boats at Needles, where this unidentified contraption ferried machinery across the river just below the Needles Reduction Company works in the mid 1890s.

1900, when she finally commenced work under Captain John W. Babson, a retired railroad man.[29]

The very existence of a rival steamer, however, prompted Polhamus and Mellon to look more critically at their own aging boats, the twenty-seven-year-old *Gila* and the twenty-three-year-old *Mohave*. The *Gila* had the better engine, but her hull required constant repair; therefore, at a cost of $27,000 they built a whole new boat for her machinery. Named the *Cochan,* she was launched on 8 November 1899. She was slightly shorter than her predecessor, had a much lighter draft, being 135 feet long, with a 31-foot beam, and drawing only 11 inches of water light. Loaded, she carried 125 tons on just under 2 feet of water. Her design incorporated nearly half a century of experience on the Colorado, and Mellon proudly announced: "She is not much for beauty but she can wax the sand bars." The *Cochan* went to work soon after New Year's Day in the new century, carrying freight for the Searchlight mines. About the same time Mellon ran the old *Mohave* into Jaeger's Slough where he left her to rot. The following year Polhamus and Mellon sold the gasoline boat *Aztec* and retired their old barges, replacing them with a massive new one, the *Silas J. Lewis*—150 feet long with a 32-foot beam. A couple enterprising Needles barbers, Charles P. and William F. Lamar, purchased the *Aztec.* They completely rebuilt her with a 70- by 12.5-foot hull of much lighter draft and went into competition for the booming upper river business.[30]

In the meantime a fast-talking franchise seeker, who styled himself Captain Alphonso B. Smith, was floating a whole conglomerate of rival steam and railroad companies to serve the Colorado River and the Gulf. Smith had first alighted in Yuma in the fall of 1893, talking the city council into giving him the sewage and streetcar franchises. He failed to act upon these, but four years later he was back with an even more "valuable" franchise from the Mexican government for a line of seagoing steamers to run between the gulf ports up to San Jorge Bay, seventy-five miles below the mouth of the Colorado, and a railroad from there to Yuma. He floated the Mexican Coast Steamship Company and envisioned a fleet of magnificent steamships, which he had already named the *Porfirio Diaz, Acapulco, Mazatlan* and *Tehuantepec*. They were to carry a hundred passengers each to a grand "tourist and health-seekers' hotel of the Mexican style," which he intended to build on the shore of San Jorge Bay. But again nothing came of the great seagoing fleet as Smith whiled away the next few years, surveying railroad lines from Yuma to San Jorge and trying to promote some mines in the hills behind.[31]

Early in 1901, however, Smith abandoned the railroad scheme and decided to run a line of steamers down from Yuma instead. Quickly hustling up $25,000 backing from two Chicago businessmen, William S. Twogood and Erven E. Busby, he formed the Mexican-Colorado Navigation Company. Smith announced that he would start twice-a-week steamboat service from Yuma to San Jorge on 1 May and that two small propellor-driven steamers were being shipped from Chicago. These two boats, one 38 feet long with a 9-foot beam and an 18-inch draft, and the other 48 by 12 feet with 34-inch draft, were to be named

The *St. Vallier,* built at Needles in 1899 by the Santa Ana Mining Company, was the first rival steamboat on the Colorado in over a third of a century, but litigation kept her tied up until the summer of 1900 when she began operating under Captain John Babson (on the upper deck right).

In response to the revival of trade Mellon and Polhamus completely rebuilt the *Gila* in 1899, renaming her the *Cochan,* and brought her up to Needles.

Passengers on the steamers found little to do but while away the time in idle conversation.

the *San Jorge* and the *Quetovac*. In addition Smith said he would build an 81-foot stern-wheeler, the *Sonora,* capable of carrying thirty first class passengers and twenty-five tons of cargo. Only the *San Jorge* ever materialized. On 3 June 1901 she set off on her maiden voyage to the gulf and the local press once again proudly hailed Yuma as the "Seaport of Arizona."[32]

To Captain Smith's chagrin, however, the *San Jorge* ran out of coal on her return trip, and because she had to wait for a sailboat to bring her fuel, she didn't get back to Yuma for two weeks. Smith also learned that a propeller was ill-suited for the shallow channels of the Colorado. Thus the following month he sent the *San Jorge* down to run exclusively on the gulf and he bought a paddle-wheel steam launch, the *Retta,* for the river run. The *Retta* was an open-decked boat, about 36 feet long with a 6-foot beam and topped with a scalloped canopy. She had been built at Yuma the previous year by Captain Frank Friant for short pleasure excursions up and down the river. Smith immediately put the *Retta* to work carrying freight up to Ehrenberg and way ports with his brother Charles M. Smith as captain. She did a brisk business, and by fall she was carrying rails up to Picacho for the California King's narrow-gauge.[33]

The *Retta's* success encouraged Alphonso Smith and his backers to get a larger boat, so in December 1901 they purchased the stern-wheel steamer *St. Vallier* from the Santa Ana company's creditors and added four cabins on her upper deck to compete for the passenger business. Under the command of Captain Joaquin Mendez, a former barge captain for Polhamus and Mellon, the *St. Vallier* commenced regular trips from Yuma to Needles in direct competition with the *Cochan*. The planned developments at San Jorge Bay were soon forgotten as the Mexican-Colorado Navigation Company fought a hot, but ultimately successful, battle for a major share of the burgeoning river trade.[34]

In the meantime yet another rival entered the field for a share of the business. In the fall of 1902 the Colorado River Transportation Company was organized by F. L. Hawley of Needles and F. L. Forrester, formerly of the Mexican-Colorado company. With a crew of Columbia River ship carpenters, they quickly built a "handsome and staunch" stern-wheel steamer, the *Searchlight*. Launched at Needles in December 1902, she was the last steamboat to be put on the Colorado River below the Grand Canyon. She was 91 feet long, with an 18-foot beam and a capacity of 60 tons, and she was fitted with six staterooms, a smoking room and a galley to serve passengers in style. Her marine boiler and 100-horsepower steam engine were installed early the following year. With Captain Hawley at the helm the *Searchlight* made her maiden voyage on 29 March 1903. This was followed by a couple inaugural excursions to Fort Mohave, one chartered by the Mohave Indians. She then began regular service from Needles to Quartette Landing, sixty miles upriver, where she made connections with the Quartette Mining Company's railroad which carried her passengers and freight to Searchlight. The *Searchlight* soon got most of the heavy trade above Needles, further cutting into Polhamus and Mellon's business.[35]

Another rival steamboat, the *Searchlight,* was built by F. L. Hawley of Needles in 1902 to compete for river business. She began regular service in 1903 from Needles to Quartette Landing where she made connections with the Quartette Mining Company's short line railroad which carried passengers and freight on up to the mining camp of Searchlight.

As if competition for trade were not stiff enough already among the steamers, countless new gasoline boats also entered the field, each taking yet another bite of the river trade. At Needles there was F. D. Spalding's 34-foot gasoline launch built in 1901; the Wright and Lawrence Mining Company's sternwheeler, launched in January 1903 to supply their mines downriver; Lichfield and Leland's 28-foot curiosity, completed in February 1903, looking something like an army tank with her paddles mounted on a sprocket chain running a third of her length; V. F. Layton's little *Mohave* (III) which sank in April 1905 on her maiden voyage downriver with equipment for the railroad surveyors at Parker; and the *Water Pearl,* a speedy, lean 32-foot triple-screw, powered by a 25-horsepower Brennan automobile motor, launched in October 1905 by the Lamar brothers and run by William Sweeney as a ferry for several years. At Yuma there was the little *Katy Lloyd,* which O. N. Lloyd tinkered with for a couple years before she sank on her first trip upriver in October 1904, and Sam Wilson's more successful *Bessie May* which started hauling light freight to Ehrenberg in August 1906.[36]

Under the aggressive competition of Alphonso Smith's Mexican-Colorado Navigation Company and a host of lesser rivals, Polhamus and Mellon gained little from the new mining boom. Thus in 1904 Isaac Polhamus retired from the river business. After nearly half a century on the Colorado, starting as a pilot on the *General Jesup,* Polhamus at the age of seventy-six announced that he had "earned a rest." Since he remained active in other business for another decade, however, it is quite likely that he simply saw the impending end of the river trade and sold out while he still had a chance. His half interest in the *Cochan* was purchased by two Yuma merchants, John Gandolfo and Joe Thornton. They, with Jack Mellon, who was still in his early sixties, kept the *Cochan* running for another five years, but a major part of that time they were operating under contract to the Mexican-Colorado company.[37]

Despite the success of his company, Alphonso Smith may also have seen that the time was ripe to sell, for no sooner had he gotten his company a dominant share of the river business than he, too, sold out. His partners bought his interest and promptly reincorporated the company in August 1904, ballooning its capital from $25,000 to $250,000, apparently so that they could push its stock onto the public at $100 a share. William Twogood became president of the company after Smith's departure, but Smith's brother Charles stayed on as captain of the *St. Vallier* until the spring of 1905 when he quit to go into ranching. John A. Anderson replaced him as captain but was quickly busted to pilot after he let both the steamer and her barge catch fire. The company's secretary, A. R. Coonradt, then took the helm. Early in 1905 the company sold their little launch, the *Retta,* to F. C. Robie and William Hutt of Needles. They completely overhauled her and put her into the freighting business between Mellen and the mines around Bill Williams' Fork. But on 24 February, her hull was ripped open by a submerged log and she sank—a total loss.[38]

In 1905 the Arizona and California Railroad reached the Colorado at Parker, bisecting the steamers' last profitable

The place of the steamers upriver was taken by a couple of smaller gasoline-powered stern-wheelers, the *Iola* and the *Hercules,* both built in 1906 by the Hall brothers. Competing for business between the railroad crossings at Topock and Parker, both, however, advertised themselves as "steamers" in the Needles newspaper.

New Stern-Wheel Steamer
"IOLA"

Is now running between Topock and Parker.

Dates of departure and other information may be obtained from

B. L. Vaughn, Real Estate and Mining Broker,

Needles, California, or from

C. S. HALL, MELLEN, ARIZONA.

THE NEEDLES NAVIGATION CO.

A General Freight and Passenger Business on the Colorado River

Steamer HERCULES makes regular trips between Topock and Parker

RATES REASONABLE

CALDWELL & TUNGATE, Proprietors.

Main Office: 400-401 Currier Building, Los Angeles, Cal.

Or Address: Care of Capt. Frank Cook, Mellen, Ariz.

stretch of river between Yuma and Needles, and the San Pedro, Los Angeles and Salt Lake Railroad passed just north of the upper reach of the river within an easy haul of Eldorado Canyon and Searchlight. By the end of the year all three steamers, the *Cochan, St. Vallier* and *Searchlight,* were taken down to Yuma where they still found some business with the booming irrigation projects. Their places upriver were filled by a couple medium-sized, gasoline-powered paddle-wheelers which carried light freight on the remaining short runs until they, in turn, were replaced by trucks.

The first of these new gasoline boats was C. S. Hall's *Iola* launched at Needles in the spring of 1906. She was a 33.5-foot long stern-wheeler with an 11-foot beam, and she was powered by a 40-horsepower engine. The editor of the *Needles Eye* heralded her as "the most successful small boat which has ever stemmed the current of the muddy Colorado." Hall put her to work in June transferring freight from Needles to the Mohave and Milltown railroad landing just across the river, but despite the views of the *Eye,* he soon became dissatisfied with her design. Moreover, his brother F. W. Hall had, in the meantime, been building a rival, larger and more powerful boat, the *Hercules,* which he launched in October 1906. The *Hercules* was a 44.5- by 9-foot side-wheeler with a 45-horsepower gasoline engine. So as not to be outdone by his brother, C. S. Hall began a new boat, the *Iola* (II), which was larger and more powerful than the *Hercules.* She was a 47- by 11-foot stern-wheeler with a 56-horsepower gasoline engine. The new *Iola* was launched on 26 January 1907, and her namesake was abandoned two months later. Hall put her to work between the railroad crossings at Mellen and Parker, while his brother sold the *Hercules,* remodeled as a stern-wheeler, to the Needles Navigation Company, organized by H. J. Caldwell and D. W. Tungate of Los Angeles. The new owners put her on the run above Needles.[39]

In addition to the fleet of gasoline boats, the turn-of-the-century mining boom brought to the Colorado a few awkward new steamcraft—the gold dredges—which were expected to become even more profitable than the graceful stern-wheelers. Profitable working of placer deposits by dredges was first demonstrated in the West in 1894 and within half a dozen years seventy-five dredges were at work throughout the West. In 1896 dredges were put on the upper Colorado basin near Breckenridge, Colorado, on a small stream called the Swan—a tributary of a tributary of the Colorado. By the turn of the century they were being introduced on the Colorado itself.[40]

Early in 1900 at least four companies, The Advance Gold Dredging Company, the Cuchan Gold Mining, Milling and Dredging Company, the Kansas City Gold Dredging Company and the Yuma-Colorado River Gold Dredging Company, were all organized to work placer ground on the river above Yuma. The Advance company began construction of the first dredge at Yuma in May under its superintendent, Henry Linn. The dredge was designed by W. T. Urie of Kansas City. Her hull was 30 by 110 feet, and she was equipped with a continuous chain of half-ton buckets, powered by a pair of 80-horsepower locomotive engines. The buckets could scoop up 4,000 cubic yards of gravel a day. The gravel was then dumped into a revolving cylindrical "grizzly," mounted midship, which removed the finer gold-bearing gravels, which in turn were run through an 80-foot sluice to remove the gold. Though the dredge cost $50,000, required a crew of fifteen, and consumed 240 cords of wood a month, the placer gravels she was built to work were said to contain as much as fifty cents in gold per cubic yard, so her owners looked forward to "immense profits."[41]

The *Advance* was launched on 23 July 1900, and six weeks later, after her machinery was installed, she headed upriver to begin work at the Pot Holes, fourteen miles above Yuma. Before she could reach the placer ground, however, she had to cut her way through an extensive sandbar. This turned out to be much slower work than expected, and she was stranded for several months during low water. Finally in May 1901, she reached the gold-bearing gravel banks, but to Linn's disappointment, she was unable to recover the gold—it was simply too fine to settle in the sluice. A couple months of remodeling failed to improve the situation and Linn suspended the operation. The following year a new firm, the Bullion Bar Dredging Company, leased the boat for further tests, but these, too, were pronounced a failure and the boat and machinery were abandoned. As the other dredging companies had cautiously awaited the results of the Advance dredge before building their own, the failure of the *Advance* brought an end to gold dredging around Yuma.[42]

Farther upriver, however, the gold-bearing bars around Eldorado Canyon lured others to attempt dredging several years later. In March 1909, a new Colorado River Dredging Company promoted by one H. J. Meyers of Detroit, commenced construction of a DuBois dredge opposite the mouth of Eldorado Canyon. Unlike the *Advance*, this new "Ship of Gold," named the *North Dakota*, was designed to suck up, rather than scoop up, 3,000 cubic yards of gold-bearing gravel a day with a large suction hose, pumped by a 100-horsepower steam engine and guided underwater by three men in diving suits. The *North Dakota* was 100 feet long with a 45-foot beam, drew 18 inches of water and cost an extraordinary $165,000.[43]

In June 1909 she began work on gravel that was supposed to average $1.25 a ton, but the runs again yielded only a trace of gold. As at the Pot Holes what gold the gravel did contain was flour gold, too fine to be recovered. The operation was finally suspended in November, and on 2 January 1910, while the company was pondering a new recovery process, the *North Dakota* swamped and sank in a flood surge on the river. The Searchlight editor concluded wryly, "The monster barge seems to have solved the problem confronting the unfortunate stockholders of what to do with the 'white elephant' by committing suicide."[44]

The failure of these two boats, the *Advance* and the *North Dakota*, marked the end of gold dredging on the Colorado below the Grand Canyon, but not an end to dredging entirely, for more

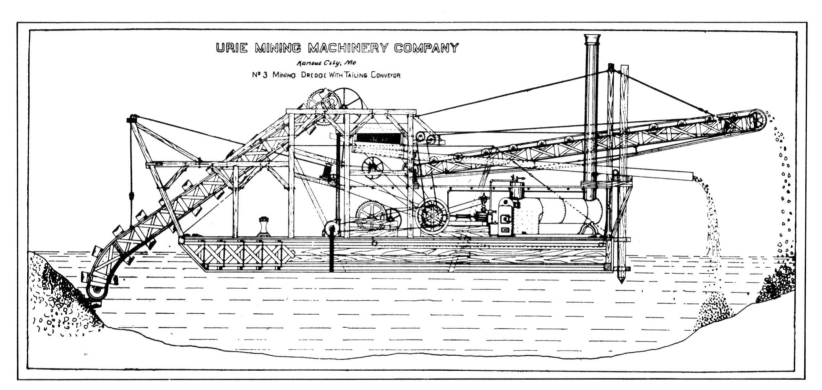

The mining revival also brought a new kind of steamer—the gold dredge—to the Colorado, but the *Advance,* built by Urie Mining Machinery Company and launched in 1900, and her followers all failed to retrieve the finely milled gold that lined the bed of the Colorado.

By the end of 1905 all of the steamers were stationed at Yuma and on this day only the *Searchlight* seems to have found work, while the *St. Vallier,* the *Cochan* and her barge *Silas J. Lewis* lie tied up at the bank at Yuma just below the ferry.

traditional types of dredges were already coming into use on the lower river. Use of dredges to cut through the bars and deepen the shallow channels to facilitate navigation had been proposed for many years. Ironically, however, the dredges that were finally introduced came not to improve navigation but to cut irrigation canals which would draw water away from the river, thereby aggravating the problem and leading ultimately to the closing of navigation. But that story will be told in the last chapter.

Thus after a quarter century of doldrums the Colorado steamboat business was booming once again and old-timers looked forward to a return of the brisk trade of those "good old boating days." This boom, however, only proved to be the final flash of a dying era. In less than a decade the business would be dead. Yet in that final flash there was a fever of activity that swept even into the deep canyons of the upper Colorado and carried steamboats far beyond the cherished "head of navigation" clear into Wyoming.[45]

The little *Major Powell,* launched at Green River, Utah, in 1891, was the first steamboat in the canyons. She made only three trips down the Green, the last of which in April 1893 was publicized by Lute H. Johnson, posing here at Wheeler's Ranch.

Steamboats in the Canyons

Steamboat Adams first promoted the idea of running steamers into the deep canyons of the upper Colorado, far above the accepted head of navigation, but the subsequent explorations of Major John Wesley Powell discouraged such notions. Still, the idea fired the imagination of romantic journalists and whetted the appetite of a few eager entrepreneurs. "Imagine the caliope" of a steamboat, one enthused, "piping its stentorian music through the canyons and labyrinths of this most beautiful and majestic scenic route on a moonlight night. The Colorado River Canyon country will be the Mecca of the world's wonders . . . and billions of dollars will be spent by the travelling public for no other purpose than the novelty of its scenes." It was the lure of these dollars that finally put steamboats into the canyons.[1]

The eager promoters, however, found the upper Colorado an inhospitable host. Half a dozen different steamboats of varying shapes and sizes were tested for fully twenty years on over seven hundred miles of the upper Colorado and Green rivers—all the way from Lee's Ferry, Arizona, to Green River, Wyoming, sixteen hundred miles by water from the Gulf of California. Each and every boat was a failure. Few even made more than a couple trips, but they still left a fascinating story in their wakes.

The vast canyon country of the upper Colorado, embracing the eastern half of Utah and adjacent portions of Arizona, Colorado and Wyoming, was settled much more slowly than the lower reaches of the river. Though the Mormons sent a mission into the canyon country as early as 1855 in an unsuccessful attempt to raise cotton, it was not until the coming of the railroads that the first permanent settlements were made. In 1868 the Union Pacific Railroad established a station at Green River, Wyoming, and the following year Major Powell began his explorations of the canyon country from that point. The Rio Grande Western Railroad penetrated the canyon country farther south in 1883, establishing a station at Green River, Utah. A few years before, Mormon farmers had begun to settle on the upper Colorado, then known as the Grand above its junction with the Green. In 1882 the town of Moab was laid out in Grand Valley, just a mile from the river and thirty-five miles from the new Rio Grande railroad. Each of these towns took an interest in river navigation, but it was Green River, Utah, that became the center of the fledgling steamboat business.[2]

The first steamboat in the canyon country had its inception in the fall of 1890. Earlier that year B. S. Ross of Rawlins, Wyoming, had taken a rowboat down the Green River to the first cataract on the Colorado, four miles below the junction with the Green. He was awed by the canyon scenery, but he was even more impressed by the possibility of exploiting it. Ross, a thirty-six-year-old, former railroad worker just recently elected to county office, was rapidly expanding his horizons. Joined by Wyoming Supreme Court Justice Homer Merrell and a few other friends, Ross floated the Green Grand & Colorado River Navigation Company to run a line of excursion steamers down the Green to a hotel they planned to build at the first cataract. They purchased a small steam launch in Chicago in the summer of 1891 and had her shipped by rail to Green River, Utah. There she was launched in mid-August and christened the *Major Powell*. She was an open-deck boat, 35 feet long with an 8-foot beam. Covered with a canvas canopy, she resembled the *Retta* on the lower river, except that her two 6-horsepower steam engines turned twin screws rather than a paddle wheel which the shallow waters required. This was not, however, the only error in her design. Ross's specifications had called for a boat which drew no more than twenty inches of water when loaded with three tons. When the *Major Powell* was floated, she took twenty-six inches empty and required a mountain of coal to feed her boiler.[3]

Ross had organized quite a party, including two journalists, to join him on the first steamer trip into what he freely termed the "Grand Canyon." But on his trial run just below the town, the *Major Powell* smashed both her propellers on a rocky bar. Replacements had to be sent from the east, and the river was falling fast; so Ross and his party went on in row boats, leaving the first steamer excursion for the following year. Ross was no longer as confident of success as he had been. On the second attempt he hired a local rancher, Arthur Wheeler, to try taking the steamer down to the first cataract, and he did not invite the press.

After putting heavy iron shields around the propellers for protection against rocks, Wheeler fired up the boiler and headed the *Major Powell* into the canyons on 15 April 1892. With Wheeler were H. J. Hogan, engineer, W. A. Heath, an artist, and Daniel Kenty, Ross's first official "excursionist." The river was greatly swollen, approaching the spring flood, so despite the added weight of the shields, the little steamer had little trouble clearing the bars. They were hung up briefly only at the Auger riffle five miles below Green River. Twenty-two miles downriver, Wheeler stopped at his ranch to show his passengers a "natural soda fountain" that bubbled up in the middle of the river. Wheeler's ranch, settled by him and his brothers in 1884, was the last settlement on the river. Just below the ranch the *Major Powell* entered Labyrinth Canyon. High walls closed in the river,

An unlikely port, Green River, Utah, at the Denver and Rio Grande railroad crossing was the center of steamboat activity in the canyon country at the turn of the century.

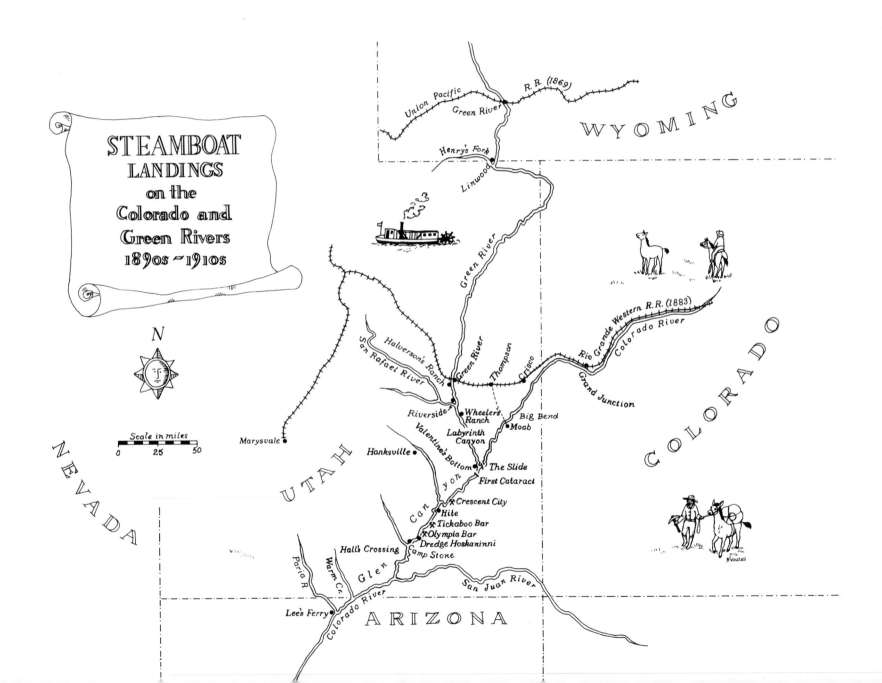

and the "picturesque part of the journey" began. Wind and water had worn the red sandstone cliffs into fascinating shapes, and mineral stains had painted them in various hues, but the greatest attraction for the steamer excursionists was the prehistoric cliff dwellings tucked in niches in the canyon wall. At some of the more accessible ruins they gathered the broken pottery, arrowheads and corncobs of a vanished civilization.[4]

Passing through Labyrinth and Stillwater canyons without incident, the *Major Powell* entered the Colorado and ran down four miles to the first cataract—120 miles below Green River, Utah. This marked the lower limit of navigation on this portion of the river, but standing at the head of the cataracts eager entrepreneurs dreamed of building a short railroad over the rapids to connect with yet another steamer on the smooth waters of Glen Canyon below. For the time being, however, the excursionists contented themselves with exploring the proposed site of Ross's grand hotel. Then they headed back upriver, putting the *Major Powell* to her first real test. Her engines were barely able to pull her against the current, so her progress was exceedingly slow. By the time she reached Wheeler's ranch her coal was nearly exhausted and they tied her up. Returning overland to Green River, they jubilantly proclaimed their successful steam navigation of the upper river. Ross now eagerly publicized the trip as a steamboat excursion into the "Grand Canyon," and although he promised to commence regularly scheduled trips, nothing came of his promises.[5]

The scheme was revived early the following year, however, by William Hiram Edwards, a young roustabout who had previously accompanied the Stanton and Best parties down the river and who would later claim for himself the honor of being the first steam navigator of the upper river. Persuading friends in Denver to lease the *Major Powell*, Edwards and two companions, H. F. Howard, an old Lake Erie steamboat engineer, and G. M. Graham, a health-seeker from Montreal, arrived at Wheeler's ranch in early March 1893. Overhauling the boiler to burn wood, they set out with the steamer on her second trip down the Green. The river was considerably lower then than it had been when Wheeler went down, and Edwards's party took a month making the 200-mile round trip to the first cataract.[6]

Much of their time was spent either on shore, cutting cottonwood and gathering driftwood for the boiler, or knee-deep in the water pulling the *Major Powell* over the bars with block and tackle. Even so, they, too, found time to take in the scenery and poke around the cliff dwellings. They also carefully appraised the agricultural prospects of the narrow bottomlands strung along the base of the canyon. One adventuresome family, the Valentines, had just settled a patch of land only a dozen miles above the junction of the Green and Colorado and some eighty miles below Wheeler's. The Valentines—who had arrived the previous fall, built a cabin and set out a vegetable garden and fruit trees—were overjoyed at the prospect of steamboat service. Edwards, likewise, was pleased with the possibility of added business for a steamer in bringing supplies down to the ranchers and taking their produce up to market.[7]

To publicize his exploits Edwards invited Lute H. Johnson, a Denver newspaperman and photographer, to accompany him on a second trip. They started from Wheeler's ranch on 27 April 1893. The river was much higher by then and the *Major Powell* made the round trip to the first cataract and back in fourteen days, of which only four days were actually spent running the river. Johnson enthusiastically wrote up the adventure as an illustrated feature for the Sunday papers, spicing it with a narrow escape in an embryo rapid just above the first cataract. Edwards at the same time unblushingly claimed that it was he who had first "proven" the navigability of the canyon country, and he implied that Wheeler had not even taken the boat below his ranch. This canyon country candidate for an Ives-Johnson dispute never reached fulmination, however, for Wheeler never bothered responding to Edwards's claims—if in fact he ever heard of them. Despite his claims Edwards concluded that the little *Major Powell* was not really of "proper construction" for regular service, and with no one rushing forward to put up money for a new boat he abandoned the scheme, apparently content to bask in his self-proclaimed glory.[8]

In 1894 four men brought the *Major Powell* fifteen miles upriver to Halverson's ranch in Little Valley. There they scrapped her, taking out the engines and boiler and leaving the hull to be swept away with the next flood. The *Major Powell* had made only three trips in three years and logged only 630 miles on the waters of the Green and Colorado rivers, but she had won her place as the pioneer steamboat in the canyon country, and though she was a commercial failure, her very failure challenged others to try again.[9]

After the demise of the *Major Powell*, J. N. Corbin, who had worked with the Colorado promoters of the boat, came to the canyon country to start a newspaper, the *Grand Valley Times*, at Moab. With a small flood of editorials Corbin kept alive the idea of steam navigation in the canyons, extolling the manifold benefits that a steamer would surely bring to farmer, merchant and tourist. His persistence was finally rewarded in 1901.[10]

In the fall of that year the second canyon steamer, the *Undine*, was launched at Green River, Utah. A flat-bottomed stern-wheeler, she was much better suited to the river and looked like a scaled-down model of the time-tested boats on the lower river. Unfortunately, however, she was scaled down in power, too, and with a captain who was no match to the hazards of the river she would come to a tragic end.

The captain and owner of the *Undine* was Frank H. Summeril of Denver. He had her built at Rock Island, Illinois, and shipped by rail to Green River. She was a coal burner with a 20-horsepower engine, was 60 feet long, with a 10-foot beam, drew only 12 inches light and could carry 15 tons on 20 inches of water. Summeril was much more ambitious than the *Major Powell's* promoters. He, too, expected the tourist trade from Green River to the cataracts to be the prime source of income, but he was also determined to try opening a shipping business on up the Colorado, or Grand, to Moab. Thus on 22 November 1901, the *Undine* set out on her maiden voyage, not just to test the Green at much lower water than the *Major Powell* had tried, but to explore the navigability of the upper Colorado. Accompanying

The *Undine*, the second and most successful of the canyon steamers, was launched at Green River in 1901 by Frank Summeril, who took her all the way down the Green and up the Colorado to Moab.

Captain Summeril were his six-year-old son, Stanley, a photographer, and a crew of four, including one of the Wheeler brothers.[11]

The voyage got off to an ominous start. A short distance below Green River the *Undine* struck a boulder midstream, knocking a hole in her bottom. She was soon patched and on her way, but she ran aground frequently. In spite of her lighter draft the water was so low that she stuck on nearly every bar and had to be pulled free with block and tackle. Nonetheless, they finally reached the first cataract on the Colorado. Here they camped for a week while Summeril looked for a suitable site to build a health resort. Then they headed up the Colorado for Moab. To their surprise it had fewer bars and proved easier to ascend than the Green. The only hazard to navigation was the Slide, just a mile and a half above the junction with the Green. It was an enormous landslide which had fallen from the north wall of the canyon and blocked fully three-fourths of the channel. At flood stage the current around it was too swift to pass, but at lower water the *Undine* had little trouble getting by. They reached Moab on 9 December, having ascended the sixty-five miles from the junction in three days.[12]

To editor Corbin the arrival of the *Undine* at Moab was a dream come true, and he hailed the venture as "the greatest enterprise that has ever been started in this section." Summeril, apparently giddy with success, claimed that he could make the 180-mile run back up to Green River in just 18 hours!—quite a trick considering it had taken him 18 days coming down. But of more direct interest to the citizens of Moab was his announcement that if they would guarantee him all their freight he would ship it to the railroad for only six dollars a ton, twenty-five percent less than the teamsters charged for the overland haul. Corbin prayed the Moabites were "alive enough to their own interests" to accept the offer. While they pondered, Summeril returned to Denver for the Christmas holidays to complete arrangements.[13]

Most of the farmers were reluctant to risk their crop on the river till the steamer was a little better tested. By the time Summeril returned he, too, had had second thoughts at least about his 18-hour claim, and when he headed the *Undine* back down the Colorado in February 1902, he expected to take about ten days making the round trip to Green River; he was going to prospect some on the way, of course. It, in fact, took him even longer, and by the time he got back to Moab he had concluded that the quickest route to the railroad would be to go on up the Colorado toward Cisco, some thirty miles above Moab. There were a number of rapids on this stretch, however, so he tied up the *Undine* for a couple months while he blasted out boulders at several points.[14]

Finally on 8 May Summeril confidently headed the *Undine* up the river. She got only about six miles to the first riffle, where her capstan line snapped while trying to pull her through. Summeril returned to Moab for new lines and started out once more only to have the lines snap again. With much stronger lines he set out on what proved to be his last trip on 21 May. This time he got the *Undine* through the first riffle, but late that afternoon she was stopped in a second at Big Bend just two miles beyond. As he tried to pull her through with the capstan, her bow caught in the current and she capsized! Summeril shouted to the engineer and fireman to save themselves, and he jumped for his life. The steamer rolled sideways down the river, losing her pilothouse and boiler. The crewmen had to scramble to hang onto the hull till she finally caught on a rock. Summeril, grabbing hold of a mattress thrown from the wreck, drifted downstream a mile and a half before he could get ashore. With bare and bleeding feet and thoroughly exhausted he finally reached a placer camp several miles below. Rescuers brought a rowboat from Moab to take the stranded engineer and fireman off the wreck.[15]

Summeril returned to Denver, promising to raise money for a new boat, but a year later he was still trying. Corbin, who but a short time before could not find kind words enough for Summeril, now came down hard on his fallen hero, dismissing him as "unmechanical and inexperienced," and claiming he had "put a kittle in his boat unfit to boil plain water in." Corbin still had faith in the ultimate triumph of canyon steamboating, however, and he soon found others willing to take on the challenge. Even so, for all their efforts the little *Undine* proved to be the only steamboat ever to reach Moab.[16]

Undaunted by the failures of his predecessors, another aspiring steamboat entrepreneur, John J. Lumsden, launched a new venture, the Green-Grand River and Moab Navigation Company in the fall of 1904. The company consisted principally of Lumsden, a forty-year-old building contractor from Grand Junction, Colorado, and a Great Lakes steamer captain, H. K. Clover from Chicago. At a cost of $15,000 they built a fancy new excursion boat, which they confidently named the *City of Moab*. She was much larger and potentially more powerful than her two predecessors, but she was actually no better designed. Her hull was 60 feet long and she had a 20-foot beam, but she drew 14 inches light, and though she was powered by two 30-horsepower Watkins marine gasoline engines, they turned twin tunnel propellers rather than paddle wheels, which the shallow water required. Lumsden seemed primarily concerned with providing attractive accomodations for his prospective excursionists. A dozen comfortable staterooms, a cozy saloon and a bath—all lit by electricity—crowded her decks, giving her more the appearance of a lazy houseboat than a fast water steamer.[17]

In a gala celebration the *City of Moab* was launched at Green River, Utah, on May Day 1905. She immediately got hung up against a piling of the railroad bridge, and it took her embarrassed captain and crew a couple days to get her free. On 12 May Captain Lumsden with four friends from Grand Junction and two crewmen finally set off on her maiden voyage down Green River for Moab. As one later recalled, "It was a foolhardy undertaking the way we started out with a entirely untried boat, a green crew of landsmen or cowpunchers, and an unknown river." They quickly gained experience, however, after running aground on countless sandbars. Even with every man overboard prying, pushing, digging and tugging with bars, poles, shovels and tackle, they spent most of a day getting off a bar at the mouth of the San Rafael. With

John J. Lumsden built the gasoline-powered *City of Moab* at Green River, Utah, in the spring of 1905, but the ungainly boat failed in her first and only attempt to reach her namesake. She was turned back at the first riffle on the Colorado above the mouth of the Green.

practice they soon learned to get the boat off a bar in an hour or two, but they never learned how to keep from getting on a bar. Because the boat was nearly impossible to steer, they simply had to "ride with the current and take their chances."[18]

On reaching the Colorado, however, they had to run against the current to get to Moab. They got only two miles to the Slide. The river was much higher and swifter than when the *Undine* had gone through. Lumsden headed the *City of Moab* into the chute, but she was unable to make headway and was swept into a big whirlpool. There she spun around battered by driftwood until she was thrown against the rocky bank. She hit with such a jolt that her pilot, Charles Anderson, was thrown overboard. Anderson swam to safety, however, and the boat was not seriously damaged. Concluding that they could go no farther up the Colorado without winches, they headed back to Green River. This proved nearly as great a trial. They got stuck on numerous bars and spent two days on the one at the San Rafael, breaking the blades off a propeller before they got free. Three of the passengers quit in disgust at this point, leaving Lumsden and the rest to work the crippled craft on up the river. They got only as far as Halverson's ranch, seven miles below Green River, before they ran out of fuel.[19]

Lumsden was still confident that he could make a success of the venture, so he and Anderson spent more than a year completely rebuilding and refitting the boat, converting her to a steam-powered stern-wheeler, lengthening her hull by 10 feet, stripping off her fancy cabins to offset the added weight of her boiler, and renaming her the *Cliff Dweller*. She was launched again in November 1905, but her new engine did not arrive until the following May. Lumsden began signing up excursionists early that summer, advertising, "A NEW SPORTSMEN AND TOURIST PARADISE, A NEW WONDERLAND FOR THOSE WHO WANT TO SEE AMERICA, STEAMBOATING THROUGH THE GREAT CANYONS!"[20]

On 6 August 1906 the *Cliff Dweller,* with Charles Anderson at the wheel, headed down the Green to try once again to reach Moab, but despite remodeling she still got stuck on one sandbar after another and failed to get even as far as before. At Valentine's Bottom still a dozen miles above the river's junction with the Colorado, Anderson turned her around and headed back "because of increasing difficulties in navigation." Even then she took a week getting back up to Green River, exhausting her coal and foraging along on driftwood. Lumsden at last conceded the impracticality of the venture. In April 1907 he dismantled the boat and shipped her to Saltair on Great Salt Lake. There remodeled once more and renamed the *Vista* she finally became a successful excursion boat.[21]

By then even editor Corbin had given up trying to promote steam navigation in the canyons and, in fact, had given up his newspaper as well. There was only one man yet to be convinced. He was Harry T. Yokey, former engineer on the *City of Moab* and the *Cliff Dweller*. Though the failures of that cumbersome craft should have been particularly obvious to him, he still believed a big steamer could be run on the Green. In the spring of 1907, as

Stripping off the *City of Moab's* cumbersome superstructure and converting her to a stern-wheel steamboat, Lumsden renamed his boat the *Cliff Dweller* in 1906 and turned to the tourist trade. She proved no more successful, however, so she was taken to Great Salt Lake the following year.

the *Cliff Dweller* was being dismantled, he began construction of his own steamer, the *Black Eagle*. She was somewhat smaller and lighter than the previous boats, only 40 feet long with a 6-foot beam, and drawing only 7 to 8 inches, but she was driven by a tunnel screw rather than a paddle wheel. Yokey launched the *Black Eagle* at Green River in June and later that summer he took her downriver on her maiden voyage. She might, in fact, have proven to be a better boat than her predecessors had she survived, but just above Valentine's Bottom one of her boiler tubes plugged with mud and she blew up! Luckily Yokey and his crew escaped serious injury, but they wanted nothing more to do with steamboats, nor, for that matter, did anyone else at Green River.[22]

Fifteen years of experimentation with steamboats on the Green and Grand rivers had clearly shown that such large boats simply were not suited to the shallow waters and swift rapids of the canyon country. At the same time, however, some success had been found with smaller gasoline launches. The first of these, the *Wilmont*, had been completed in August 1904 by Edwin T. Wolverton. She was a 27-foot long stern-wheel boat, with a 5.5-foot beam and 10-inch draft, and she was powered by a 4-horsepower gasoline engine. Her total cost was only about $400. Wolverton built her to carry supplies to a manganese mine he was opening for the Colorado Fuel and Iron Company at Riverside, twenty-five miles below Green River. On his first trip upriver, however, he found that she was underpowered. So early the following year he put in a 7.5-horsepower engine and converted her to a side-wheeler, which also lightened her draft to 7 inches. These changes were so successful that Wolverton decided to go into the excursion business on the side. In April 1905 he took the *Wilmont* on her first excursion up the Colorado, making the 160 miles from Riverside to Moab in just thirty-one hours running time.[23]

Wolverton's success prompted Milton Oppenheimer of Green River to build a similar side-wheel launch that fall. His boat, the *Paddy Ross,* was the same length as the *Wilmont*, but was a foot wider with an inch deeper draft. She was also nearly twice as powerful with a 14-horsepower gas engine. Oppenheimer converted her to a stern-wheeler two years later, and she ran on the river for nearly a decade.[24]

Wolverton, in the meantime, built a fleet of gasoline boats and scows. Early in 1905 he became manager of the Utah-Nevada Copper Company's mine, twenty-five miles farther downriver below Riverside, and built two ore scows for the *Wilmont* to push. To ease her task he put a 14-horsepower engine in her that winter and built a new paddle-wheeler, the *Colorado*, for her old engine. In 1906 he also built a 33-foot, 14-horsepower stern-wheeler, the *Marguerite,* for Tom G. Wimmer who had bought the Wheelers' ranch. For a time Wolverton had a thriving business with the *Wilmont* and *Colorado*, taking copper ore up to Green River, Utah, and excursion parties down to the cataracts and up to Moab. In the winter of 1907–08 the *Wilmont* was caught in ice on the river and badly damaged. Since the copper mine was failing, Wolverton replaced her with a smaller, less powerful boat, the *Navajo*. She was 22 feet long, with a 5-foot beam, 18-inch draft, and she was driven by a screw propeller, powered by the 7.5-horsepower engine, which he took from the *Colorado,* converting the latter to a scow. Despite the *Navajo's* greater draft and propeller drive, Wolverton claimed to have run her without difficulty for over four years, until he quit the river in 1912.[25]

Quite a number of other gasoline launches were put on this stretch of the Green and Colorado rivers. In the spring of 1909 Henry E. Blake built a 24-foot, 14-horsepower propeller-driven boat, the *Ida B.,* which he ran successfully from Green River to Moab trying to promote a regular excursion line. Even Harry Yokey ventured back into the river business that same year with a little 6-horsepower boat which he sentimentally dubbed the *Baby Black Eagle.* Many others followed, the largest of which was the Moab Garage Company's stern-wheeler "The Big Boat" built in January 1925 to haul oil drilling equipment down the Colorado for the Mid-West Exploration Company. This was a big open-deck boat, 75 feet long, with a 16-foot beam, a draft of only 4 inches light, and a 40-horsepower automobile engine. Clarence Baldwin ran this and a couple smaller boats until drilling ceased in 1927.[26]

It is a curious irony that no sooner had the citizens of Green River, Utah, given up the idea of putting a steamboat on the river than the citizens of Green River, Wyoming, nearly four hundred miles upstream picked it up. Since Major John Wesley Powell first set off down the river from the Wyoming town in 1869, only rowboats had followed for nearly forty years. But in the spring of 1908 Marius N. Larsen set about building a steamboat. Larsen ran a general store in Linwood, Utah, a small isolated farming community on Henry's Fork, some ninety miles below Green River, Wyoming, and he looked to a steamboat to cut freighting costs both on goods brought in and on produce shipped out. In partnership with several Green River and Linwood businessmen, Larsen organized the Green River Navigation Company in March for the ambitious purpose of running passenger and freight service all the way from Green River, Wyoming, to Green River, Utah. Construction of the steamer went swiftly in the hands of Larsen's brother, Holger, who had worked in the shipyards in Germany. She was a 60-foot stern-wheeler with 12-foot beam, a 60-horsepower boiler and two 20-horsepower engines.[27]

Christened the *Comet* by Maurius's daughter Beulah, her launching was the main event of the Fourth of July celebration in 1908. Three days later with the call of "All aboard for Linwood!" the *Comet* headed down the Green on her maiden voyage. Holger Larsen was at the wheel, Marius was honorary purser and some of her other stockholders filled in as crew. The river was high and the *Comet* made quick time going down, reaching Linwood in less than eight hours. Her future success seemed assured, but then came her return trip back upstream. Like her predecessors in the canyons below, she was almost constantly aground on one bar after another. Finally she ran out of coal and more had to be

The oil boom prompted Clarence Baldwin of the Moab Garage to build this gasoline-powered stern-wheeler known only as the "Big Boat" to haul drilling equipment down the Colorado in 1925.

No sooner had the citizens of Green River, Utah, given up the idea of putting steamboats on the river than those of Green River, Wyoming, took it up. Their steamer landing was on the right just below the Union Pacific railroad bridge.

The stern-wheel steamer *Comet,* seen here with her construction crew, was built at Green River, Wyoming, in the spring of 1908 by Marius Larsen to carry goods to his store at Linwood, Utah, ninety miles downriver.

brought down by pack horses. It was many days before she got back up to Green River again. Still her owners confidently pointed out that she had only taken thirty-three hours "actual running" time, and with coaling stations set up along the way, that time should be improved. But the *Comet* apparently made only one more trip to Linwood, carrying the winter stock for Larsen's store. As the water was lower then and the boat more heavily laden, she ran aground many times even going down. Each time the passengers and crew had to unload the cargo to get her over the bars. It was a long, tedious trip that no one wished to repeat.[28]

After this fiasco the *Comet* was tied up at Green River and used occasionally as a local excursion boat. Even in that she lost out to a couple smaller gasoline launches, the *Teddy R* and the *Sunbeam,* which had been put on the river at the same time. Thus her engines were finally removed and her hull was left to rot below the highway bridge. The only reminder of her existence was her ship's bell which did service for many years in the local Lions Club, until it was lost in a fire.[29]

Despite the failure of five separate steamboats, one more steamer was still to be tested in the canyons—this time 700 miles below Green River, Wyoming, on the placid waters of Glen Canyon just above the Grand Canyon. The sequence of things here, however, was completely reversed from that elsewhere on the river; the steamboat, always the first powered boat to be tried, was the last here, preceded by both the fleet gasoline launch and that leviathan, the gold dredge. It was gold, in fact, that brought men and boats into Glen Canyon. Cass Hite first found rich placers on Tickaboo Bar near the upper end of the canyon in the late 1880s. By the early 1890s several hundred men were panning and sluicing for gold at every bar for fifty miles along the river from Crescent "City" above Hite's Ferry down to Hall's Crossing.

Though most miners were barely making wages with such rudimentary mining methods, an enterprising engineer, Robert Brewster Stanton, envisioned enormous profits from a highly mechanized dredging operation. Stanton, who had first visited the placers in 1889 while surveying a railroad route through the bottom of the Grand Canyon, saw Glen Canyon as "nature's sluice box," holding an estimated $140 million in gold dust just waiting to be cleaned up. Starting in the summer of 1897 he surreptitiously located a continuous string of mining claims running the entire length of Glen Canyon from Hite to Lee's Ferry— 165 miles! With these claims as a base he found backing in the East and organized the Hoskaninni Company to put a fleet of gold dredges on the river, predicting they would turn a profit of over $1 million a year. Stanton was both vice-president of the company and superintendent of its operations.[30]

Despite the lure of such fabulous profits the project moved slowly as Stanton spent the next couple years drilling a myriad of test holes in the riverbed to determine the richest spot to put the

The discovery of gold in Glen Canyon triggered a mining rush that rivaled those on the lower Colorado and eventually led to the building of the last of the canyon steamers.

first dredge. To supply his camps he brought in an 18-foot, propeller-driven gasoline launch in May 1898, but she proved to be of questionable value. Her engine gave constant trouble and was unable to make headway against even a slight current, so she had to be pulled most of the way upriver. Stanton never bothered naming her, but his crew dubbed her the "white elephant."[31]

The first dredge site, Camp Stone, was selected on a bar four miles above Hall's Crossing, and a supply road was opened from Hanksville to the river a mile and a half below the camp. The lumber and machinery for the dredge were shipped by rail to Green River, Utah, hauled overland 100 miles to the river, and taken by barge up to the camp. The launch was unable to tow the barge even that short distance, however, so sails and poles had to suffice for power. Construction of the dredge, *Hoskaninni*, began in June 1900, and early the following year she was ready to begin work. The dredge, which cost about $25,000, was designed by the Bucyrus Company of Milwaukee. Her hull was 105 by 36 feet, and her equipment consisted of a chain of forty-six buckets for scooping up the gravel, a rotating, double-barreled grizzly for sorting out the coarse rock and a sluice box for settling the gold dust. She was powered by five separate gasoline engines, which generated a total of 168 horsepower.[32]

Dredging commenced in February 1901, but breakdowns plagued the operation. After more than a month of difficulties Stanton complained that he was "worn out with worry and disappointment," but continue he must. The outlook grew even more dismal, however, when he made his first cleanup of the

Robert B. Stanton, attracted by the gold discoveries in Glen Canyon staked claim to 165 miles of the Colorado riverbed and set out to dredge it.

Stanton's dredge, the *Hoskaninni*, was built in Glen Canyon at Camp Stone shown here, four miles above Hall's Crossing, and launched in the fall of 1900.

The *Hoskaninni* commenced operation in February 1901, but three months' work yielded less than seventy dollars worth of gold and Stanton abandoned the boat.

sluice on 13 April. He found that in two months' operation the *Hoskaninni* had recovered only $30.15 worth of gold!—barely a tenth of one percent of what he had expected. Stanton moved the dredge to another spot but to little avail. Three weeks' work there yielded only $36.80. He finally realized that the dredge was simply unable to recover the fine flour gold that his tests had shown the gravel contained—a realization that the owners of the *Advance* dredge, a thousand miles downstream, were coming to at almost exactly the same time. With many outstanding debts the Hoskaninni Company passed into receivership and the entire property, dredge and all, was sold in December 1901 for $200. A watchman, who looked after the leviathan for a couple of years, got the deed as a settlement for back wages. Little was ever salvaged from it and the decaying hulk was finally covered by the rising waters of Lake Powell.[33]

A few gasoline boats were tried in the canyon in subsequent years. Frank Bennett, manager of the Moquie Mining Company's operation on Olympia Bar upriver from Stanton's dredge, built a 28-foot stern-wheel launch, the *Lucy B*, at Hite in 1902 to help supply his camp. She proved no more useful than Stanton's launch, however, since her two cylinder automobile engine, supposed to deliver 6 horsepower, gave only half that and could not power her back up to Hite. Bennett pulled out the engine and converted her to sail. Harry Yokey built a new power boat for Bennett at Green River in 1905. She was a 22-foot, propeller-driven launch with a 12-horsepower engine. She had power enough to stem the current, but her 16-inch propeller struck bottom on the bars, so Bennett could not take her much farther upriver than her predecessor. When she sank at Tickaboo the following year, Bennett recovered her engine to try on one more boat, but this, too, was a failure, and he went back to sail.[34]

By 1910 most of the miners had left Glen Canyon to seek their fortunes elsewhere, but one diehard, Charles H. Spencer, was still determined to wrest the gold from the canyon and he had a novel, if quixotic, scheme for doing so. It was a scheme which also led to the building of the last steamboat ever put on the Colorado River. Spencer believed that the source of the placer gold was the shale and sandstone formations through which the canyon was cut, so he set out to mine the canyon walls. The easiest rock to work was the Chinle shale which outcropped at Lee's Ferry. The shale was readily broken up by water, so he set up a Rube Goldberg apparatus to hydraulic and sluice the deposit. Just as a backup operation, he also constructed a makeshift suction dredge at the same spot to try to recover the gold from the riverbed.[35]

Before Spencer could test either operation, however, he needed fuel for the boilers. Rather than pay the exorbitant costs of freighting 140 miles from the railroad, he decided to work a low-grade coal deposit on Warm Creek, 28 miles upriver, as a further expansion of his operation. To bring the coal downriver

The last steamboat in the canyon country was built to supply coal for this Rube Goldberg contraption set up at Lee's Ferry by Charley Spencer in an ill-fated attempt to extract gold from both the riverbed and the shale of the canyon walls.

he built a couple gasoline launches—the 27-foot *Violet Louise* and the smaller *Mullins*. Neither boat, however, was able to carry enough for his needs. Thus in the summer of 1911 he contracted with Schultz, Robertson and Schultz of San Francisco to build a stern-wheel steamboat, costing about $30,000. That fall she was shipped in pieces to Marysvale, Utah, and hauled by ox team more than two hundred miles to the river at the mouth of Warm Creek. There the steamer was completed in late February 1912 and christened the *Charles H. Spencer*. Measuring 92.5 feet overall, and having a 25-foot beam, she was not only the last but the largest steamer ever built in the canyons. Her size, however, was a liability. She was powered by a 100-horsepower boiler and drew 18 to 20 inches of water light.[36]

Early in March the *Charles H. Spencer*, loaded with just enough coal for her own boiler, headed down the canyon on her maiden voyage to Lee's Ferry. Peter Hanna, the only man in Spencer's crew with any riverboat experience, was at the wheel, but like the canyon boats before her the *Spencer* ran aground almost immediately. It was evening before she was afloat again, so Hanna tied her up for the night. When they set out the next morning, he proceeded more cautiously, turning the steamer stern to and backing her down the river. To further slow her descent he dragged a 100-foot log chain from her bow, but it caught between rocks and broke off. He had no further trouble running aground, however, and reached Lee's Ferry that afternoon.[37]

Charley Spencer was confident as his steamer started back upstream with a barge to bring down a load of coal. She promptly grounded on a bar, however, just above the ferry and was stuck there for three days. Moreover, when she finally reached Warm Creek, the barge got away and was lost down the river. The steamer brought down a little coal on her deck and pushed the ferryboat back up instead of the barge. With several tons of coal brought down on the ferry, Spencer at last put his scheme to a test, only to come to final disillusionment. Try as he might, he was unable to get the fine gold out of the Chinle shale and he had no better luck with the suction dredge getting it from the river bottom. Thus after two years the whole scheme collapsed and was abandoned. The *Charles H. Spencer*, having run less than one hundred fifty miles on the river, was left to rot just below the ferry. By the mid-1970s only the battered rusty boiler and some scattered timber remained of the Colorado's last steamboat.[38]

Though Glen Canyon Dam has destroyed much of the canyon country, that one romantic vision of a paddle-wheeler churning through the depths of the canyons still refuses to die. Late in 1971 a former Colorado school teacher, Tex McClatchy, began construction of a mammoth stern-wheeler, the *Canyon King*. She was a 93- by 26-foot, double-decker run by a diesel marine engine. Launched at Moab on 30 April 1972, she began making regular excursions down the Colorado in May of that year —the final realization of that persistent dream that put the first steamboats into the canyon country.[39]

Spencer built his steamboat in the fall and winter of 1911 deep in Glen Canyon at the mouth of Warm Creek.

Modestly christened the *Charles H. Spencer,* she was launched in February 1912, but she made only a few short trips hauling coal before she and the whole venture were abandoned.

Left to rot just below Lee's Ferry, the *Charles H. Spencer* was still holding together quite well when this picture was taken in 1921, but little remained half a century later.

The spirit of the steamboat days was revived in the canyons by Tex McClatchy's big diesel-powered stern-wheeler, the *Canyon King,* built at Moab in the spring of 1972.

The twilight of steamboating on the Colorado.

Closing the River

The era of steam navigation on the Colorado essentially ended with the completion of the Laguna Dam which closed the river fourteen miles above Yuma. But the trainloads of celebrants who gathered at the dam on 31 March 1909 gave little thought to its passing. Instead, stuffing themselves with barbecued beef and serenaded by the Industrial Liberty Band, they applauded local politicians who hailed the new era of agricultural prosperity the dam would bring to their "American Nile." To them the Colorado was no longer a turgid avenue of commerce, but a bountiful source of water which could make the desert bloom and enrich the pockets of all.[1]

This new era dawned on the Colorado with a reawakening to the agricultural and financial potential of irrigating its flood lands; it was the opening of these lands that led to the closing of the river to steamboats forever. Ironically one of the early actors in this final drama was Eugene S. Ives, son of Joseph Christmas Ives, who just forty years before had played a role in opening the river.

The irrigation potential of the Colorado had, in fact, attracted the attention of overland immigrants even before the commencement of steam navigation. Its realization, however, had been much slower in coming. Many a westering argonaut had been impressed by the vast network of ancient Indian canals along the Gila which clearly demonstrated the potential of even so intermittent a stream. In 1849 one visionary immigrant, Dr. Oliver M. Wozencraft, seeing the overflow waters of the Colorado surging down the Alamo channel toward the Salton Sink, dreamed of irrigating the whole Colorado Desert. His greed got in the way of his dream, however, and he spent the rest of his life vainly trying to convince the Congress to grant him all the land between the river and the Coast Range—a six-million-acre empire. Yuma ferryman L. F. J. Jaeger made a more modest, but no more successful, attempt at irrigation in 1857. He spent over $50,000 digging a ten-mile canal to bring water to the bottomlands just west of the river, only to find that it ran uphill. In the decades that followed, a seemingly endless succession of irrigation schemes was proposed for diverting the waters of the Colorado, but none ever got beyond the planning stages until just before the turn of the century.[2]

Then in 1897 Eugene S. Ives floated the forward-looking, but ill-fated, State of Arizona Improvement Company. Ives, who had come West to trade on his father's "glories," proposed a grand canal system to irrigate 160,000 acres of bottomlands from Castle Dome to the Mexican line. Unlike his predecessors, he actually started work. To dig the network of connecting ditches he got convicts from the Arizona Territorial Prison and to dig the main canal he built a monster dredge—the first on the Colorado. Ives's dredge was nearly 150 by 70 feet, drawing more than 2 feet of water, and mounted with a 4-cubic-yard steam shovel. It was an ill-designed affair which ultimately failed in the two major tasks it was put to, yet its impact on Colorado steamboating was immense, for it was these very failures that hastened the end of steam navigation on the river.[3]

The dredge was launched at Yuma in December 1897 and was towed up to Castle Dome the following month. There she proudly commenced work only to become stranded within a few weeks. By the time high water came to float her that spring, Ives had become overextended financially and he abandoned the venture. Others soon took up the idea, however, turning ultimately to the new U.S. Reclamation Service for assistance. Their pleadings finally led to the creation of the Yuma Project which envisioned not only a larger canal system throughout the bottomlands but also the massive Laguna Dam on the river.[4]

In the meantime Ives's dredge had fallen into the hands of a new group of fast-dealing promoters who had finally figured out a way to turn a buck from Wozencraft's seemingly chimerical dream of irrigating the whole Colorado Desert to the west. The federal government was no more ready to grant a fiefdom to anyone than before; indeed, the Homestead Act of 1859 and the Desert Land Act of 1877 had limited the amount of public land any individual could file upon to 320 acres—sufficient for a farm but not an empire. But an irrigation engineer named George Chaffey devised a Byzantine scheme for circumventing this well-intended legislation. Having taken a handsome profit from a couple of Southern California land promotions, only to lose most of it in an Australian scheme, Chaffey was eager to recoup his losses.[5]

Thus in the spring of 1900 Chaffey joined with another engineer, Charles Rockwood, who had followed unsuccessfully in Wozencraft's footsteps after having seen the ease with which the Colorado's flood waters had surged into the Salton Sink again in 1891 and 1892. Through a facade of dummy corporations Chaffey finally constructed a system for ripping off a fortune from the public lands which Wozencraft had coveted. Though any homesteader could buy up to 320 acres of desert land from the government for only $1.25 an acre, he could not get title until he provided water to his land. This was where Chaffey entered. He organized "mutual water companies" which turned over their

The Ives dredge, built at Yuma in 1897, played a prominent, if ill-starred, role in the closing of the Colorado, starting in 1900 when, christened the *Alpha,* she began digging the California Development Company's canal, as shown here, to bring water to Imperial Valley in the Colorado Desert.

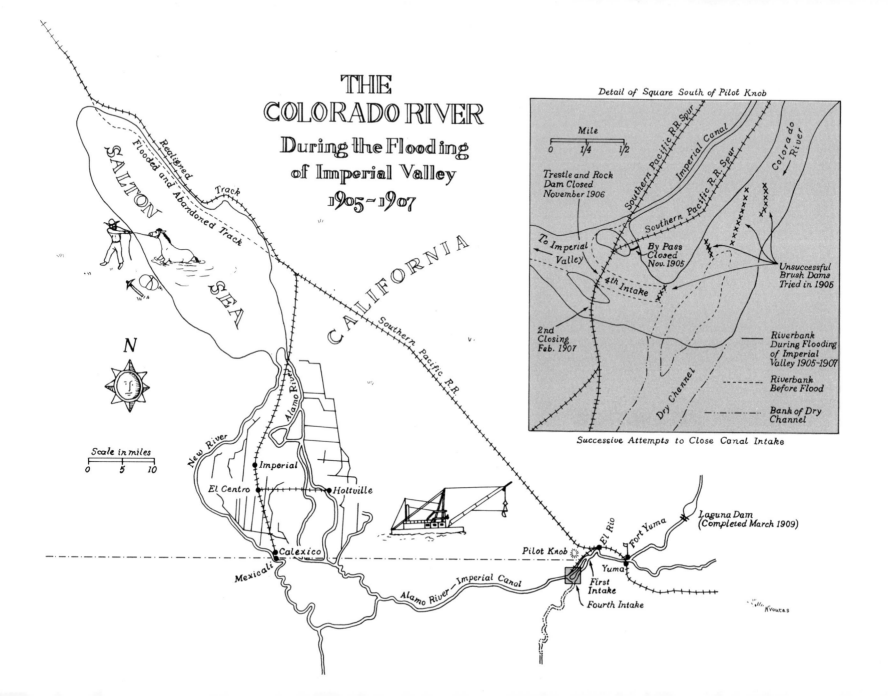

stock to a parent corporation, the California Development Company, in return for the right to buy water from the latter. The development company then posted a claim for 20,000 acre-feet of water on the bank of the Colorado and proposed to build a canal. To finance the canal it in turn sold the water company stock at a discounted rate to yet another Chaffey enterprise, the Imperial Land Company. This company, which in the beginning had essentially no land to sell, mounted a publicity campaign that brought in homesteaders by the trainload, helped them file for the government land, and sold them water rights with which to secure their title in the form of the water company stock. The stock was sold at prices ranging from about $10 to $20 a share for each acre of land claimed. Thus the land company turned enormous profits from the sale of land it never owned. Moreover since most homesteaders could not afford to pay cash for the $3,000 or more worth of water stock they had to buy, the land company took mortgages on their property, and when they failed to meet the payments on their stock, the land company soon had land to sell as well.[6]

To make the whole scheme work, of course, Chaffey had to deliver the water, but this, too, he did with as little real investment as possible. For most of his canal he simply used the Alamo channel, which had already worked quite well without him. Thus he only had to dig an eight-mile connecting canal from the channel to the river just above the Mexican line, because he had not yet gotten permission from the Mexican government to take water in Mexico to sell in the United States. If he had, he probably would not have dug any new canal at all, but would simply have deepened the natural intake to the Alamo channel several miles below the border. This, in fact, is about what the company ended up doing anyway, with disastrous results.

Purchasing the old Ives dredge, which he named the *Alpha,* Chaffey put her to work in August 1900 at Pilot Knob, less than a mile above the border. At the same time crews with horse-drawn Fresno scrapers started work at the Alamo end of the ditch. Most of the Alamo channel and the connecting canal ran through Mexico on land obtained from Guillermo Andrade by yet another dummy company, La Sociedad de Irrigación y Terrenos de la Baja California, commonly known as the Mexican Company. In less than a year the ditches and channels were all connected and the first water flowed into Imperial Valley in June 1901.[7]

By then the land company had sold more than $2 million worth of water stock and they were besieged by thousands of would-be settlers and speculators eager to buy more. Chaffey was squeezed out by his partners the following year after an unsuccessful takeover bid of his own. Still he realized $300,000 for his two years' work, and as it turned out he was lucky to get out when he did, for his hastily built canal system and his partners' shortsightedness would soon place the enterprise deeply in debt. Ironically the scheme was ruined by its own success. As unexpectedly large numbers of settlers poured in, the land company eagerly sold more water than the canal could deliver. Thus, within a short time, the development company faced possible lawsuits for breach of contract.[8]

Trainloads of settlers and speculators in the early 1900s paid the canal builders up to $20.00 an acre for water stock with which to claim public lands for $1.25 an acre, or, as shown here, bid much more for city lots in the new boom towns such as Imperial.

Fresno scrapers aided the dredge in digging connecting canals and deepening the old Alamo overflow channel.

Having transferred most of the assets to the land company, Rockwood and his partners now offered to sell the development company to the settlers, the U.S. Reclamation Service, or to anyone else who would take it, for a price, of course. The price to the settlers and the government started at $5 million. Though later cut to $3 million, the settlers could not even afford that, and the government could not hold land in Mexico. Finally, desperate even for sufficient funds to maintain token delivery of water, without which the land sales would collapse, Rockwood turned to E. H. Harriman of the Southern Pacific Railroad. The railroad, doing a profitable business from the flow of freight and produce in and out of the valley, had an obvious interest in maintaining development there, but Harriman proved to be a much tougher customer with whom to deal. In exchange for a modest $200,000 Rockwood and his partners turned over a controlling interest in the stock and management of the California Development Company and the Mexican Company to the Southern Pacific Railroad. Though Rockwood and his friends still held the profitable end of the business—the Imperial Land Company—their mismanagement of the canal soon brought calamity to the whole venture.[9]

No sooner had the muddy waters of the Colorado first been turned into the canal than they began to silt up the channel cutting the flow into the valley. Chaffey had found that the *Alpha* was unable to remove the silt as fast as it was deposited and had built a suction dredge, the *Beta,* to assist. Even their combined efforts, however, were not sufficient to keep the canal clear. When Rockwood took over he tried running the *Cochan* up and down the canal with a heavy dragline, hoping to stir up the silt enough to flush it on through, but to no avail. Finally during low water in the winter of 1902 the first and most heavily silted mile of the canal was abandoned, and a new intake was cut to the canal farther downriver. This cheap expedient of simply cutting new intakes without any headgate to control the flow became the company policy until March 1905 when disaster struck. By then the fourth intake tapped the Colorado four miles below the Mexican line and barely two miles above the point where the river had overflowed into the Alamo channel during the floods of 1891 and 1892—which had first attracted Rockwood to the scheme.[10]

The first flood in early February 1905 was much higher than those of the previous few seasons and might well have served as a reminder to Rockwood that worse could come. He was still so fearful of silting, however, that instead of preparing to reduce the flow into the intake he put the *Alpha* into it to dig it deeper. When subsequent flood crests hit within a few weeks they further deepened and widened the intake until by the first week of March the flow into the canal was out of control; the summer floods were yet to come. Rockwood, however, still failed to recognize the danger, and by the time he finally did decide to close the intake later that month he was too late with too little. The previous two seasons he had closed the intakes by dropping a makeshift brush and timber plug across the entrance, but when he tried it this time the whole mess was swept away. His concern increased as the intake grew wider and wider, more and more water poured into the canal and flooding began in the valley. In April the flood

The failure of the *Alpha* to keep the canal intake free of silt led to construction of this suction dredge, the *Beta,* but even together they were unable to remove the silt as fast as it settled.

To maintain the flow to the valley the canal company finally adopted the ultimately disastrous expedient of having the *Beta* and *Alpha* cut new intakes farther downriver without headgates for control.

receded slightly and Captain Mellon suggested closing the breach by sinking one of his barges, loaded with sandbags, in the narrowest part of the channel. Rockwood rejected Mellon's proposal, but at least one later observer, Godfrey Sykes, concluded: "There is little doubt that this plan would have succeeded had it been adopted and that the whole course of river and valley history would thereby have been changed."[11]

Instead subsequent floods further enlarged the intake until the entire river was pouring into the canal, flooding thousands of acres of newly opened farmland, eroding and scarring thousands more, washing away part of the towns of Calexico and Mexicali, forcing the abandonment of the salt works and Southern Pacific tracks, turning the Salton Sink into the Salton Sea, and leaving millions of dollars of damage in its wake. It would take nearly two years of work and struggle, and cost $3 million more before the Colorado was back in its old channel again.

After his first failure Rockwood had made a second attempt to close the intake by building a brush and piling wing dam to try to deflect the river from the intake. A pile driver was mounted on a small barge, the deck of the *Silas J. Lewis* was cleared to construct brush mats, and the steamer *St. Vallier* was chartered to tow the barges and supply the whole operation. This effort was stopped in June, however, when Rockwood decided to quit until the summer floods subsided.[12]

By June 1905, when the Southern Pacific formalized the agreement which Harriman had worked out months before for taking over the California Development Company, the company was faced with additional lawsuits as a result of the rising flood damages. With the river still unchecked and damages continuing to mount, prompt action was needed. Epes Randolph took over the presidency of the company for Harriman and appointed F. S. Edinger to replace Rockwood as chief engineer. They began a more vigorous effort to close the river but by then the problem had grown enormously.[13]

In October 1905 Edinger commenced a new diversion dam to block off the intake. He assembled a force of 300 men for the undertaking, bought the steamer *Searchlight* and leased the barge *Silas J. Lewis* for fifteen dollars a day to aid in the task. Work on the dam was progressing rapidly when on 29 November a flash flood from the Gila swept most of it away.[14]

Edinger resigned soon after, and all efforts to close the intake were stagnated as Rockwood, by default, again became the temporary chief engineer. As a result four months of the lowest water were wasted before Randolph appointed Harry T. Cory to direct the effort in April 1906, just as the spring floods began. Before these floods subsided they had further enlarged the intake until it was more than half a mile wide.[15]

Cory and his construction superintendent, Thomas J. Hind, implemented a much more solid plan for closing the break than his predecessors, and by then the Southern Pacific management was prepared to commit whatever funds were necessary to do the job. Their plan was to build a railroad trestle across the intake from which carloads of earth and rock could be dumped into the breach, rather than to rely simply on brush to trap sediment.

The spring floods in 1905 poured unchecked into the canal and half the town of Mexicali was washed away by the runaway river.

Eventually the entire Colorado left its old channel and raged through the canal flooding the Southern Pacific railroad tracks in the Salton Sink and turning it into the present Salton Sea.

The canal company put the *St. Vallier* to work in late spring of 1905 attempting to close the intake with brush mats, but the effort failed.

Work on the railroad spur from the Southern Pacific main line down to the intake began in July 1906. The tracks reached the break on August 15 and with a pile driver mounted on the end of a flatcar the trestle was slowly extended out into the current and across the breach. Nearly 1700 carloads of rock, quarried at Pilot Knob, were dumped into the channel on both sides of the trestle building up a solid dam and slowly reducing the flood into the valley. On 4 November the flow through the break was finally cut off. The Colorado once again flowed down its old channel to the gulf and the devastation of Imperial Valley was ended.[16]

Word of the closure was greeted with wild celebration in the valley. The valley also finally seemed assured of a dependable water supply for irrigation through a new concrete headgate built on the old canal just above the border, and a new, more efficient clamshell dredge, the *Delta,* had been built to keep the channel clear of silt. The new dredge, 120 feet long with a 54-foot beam, was launched at Yuma 15 August 1906. This still proved to be only a temporary solution, however, for the problem of silting was not solved until the Colorado was finally dammed.[17]

The Colorado River was not yet controlled either, for scarcely a month after the closure, while the settlers were still rejoicing, flood waters outflanked the closure dam and the whole river once again poured into the canal, threatening a repeat of the devastation. The break was so sudden that the steamer *Searchlight,* which had gone farther downriver to pick up work crews, was left stranded in the dry bed of the river, and her passengers and crew had to walk ashore.[18]

This time, however, Harriman claimed that the Southern Pacific had already spent more money closing the intake than they could ever regain from the California Development Company, and he refused to lift a hand without some new deal. The valley residents frantically turned to the government, and President Theodore Roosevelt telegraphed Harriman that it was "the imperative duty of the California Development Company to close this break at once." Harriman protested that "we are in no way interested in its stock and in no way control it," but ordered his engineers to close the break.[19]

Cory promptly launched a new offensive. Enough water was diverted back into the old channel to refloat the *Searchlight* on 28 December, and with her aid in ferrying supplies the break was finally closed again 11 February 1907.[20]

The work of returning the Colorado to its channel had cost a little more than $3 million. Whether the railroad recouped these expenditures, plus another $900,000 in damages to its tracks, is still debatable. Superficially, at least, it would appear to have done so. Southern Pacific was the principal creditor of the California Development Company when it realized $3 million from the sale of the canal system to the valley farmers; Harriman's interests also took over the land holdings of the Mexican Company for claims against it; and finally the Congress gave the railroad more than $1 million for closing the second break.[21]

All the while that the Southern Pacific had been struggling to dam the break, the U.S. Reclamation Service was engaged in an even more ambitious effort to dam the Colorado itself in order to

Harry T. Cory, who took over the task of closing the intake in 1906, built a trestle across it and brought in nearly 1700 carloads of rock with which he finally dammed the intake in November.

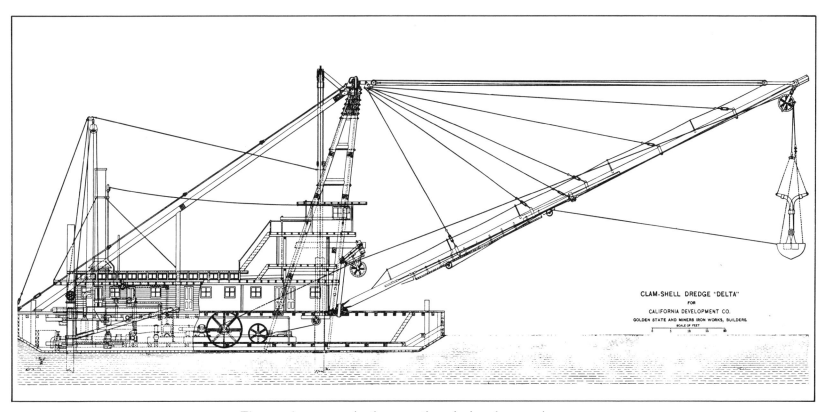

The canal company in the meantime designed a massive new dredge, the *Delta,* to try to control silting in the canal.

Launched at Yuma in August 1906, the *Delta* just squeezed through the swingspan of the Southern Pacific railroad bridge on her way downriver to the canal heading.

Barely a month after the closing of the intake, the Colorado outflanked the dam and broke back into the canal, threatening the valley with further destruction and leaving the *Searchlight* and her barge temporarily stranded in the dry channel two miles below the break.

On 11 February 1907 the break was finally closed, the Colorado was returned to her channel, and the flooding of Imperial Valley ended at last.

The receding waters in Imperial Valley revealed thousands of acres of land scarred by the man-made floods.

provide irrigation water and prevent the recurrence of such calamitous floods. The effort had begun even before the Ives dredge had cut the fateful intake on the Mexican bank. In the summer of 1902, soon after the passage of the Reclamation Act, which set aside funds from public land sales for the construction of dams and irrigation systems, the settlers along the Colorado below Yuma appealed to the U.S. Reclamation Service for aid. Surveys began that fall and in May 1904 the Secretary of the Interior allocated $3 million for construction of an irrigation system for the lower Colorado valley. The Yuma Project, as the system came to be known, ultimately provided water for 100,000 acres of bottomlands and included not only a hundred miles of irrigation canals but a massive dam across the Colorado at Laguna Landing above Yuma—a dam that sealed the fate of steam navigation on the Colorado.[22]

Bids were opened for construction of the Laguna Dam in March 1905, just as the first floods were pouring into the Mexican intake. J. G. White and Company of New York got the contract for $797,650, and work commenced on 20 July as the summer floods receded. The dam was a concrete, rock- and earth-filled structure over two hundred feet thick, up to forty feet high and extending for nearly a mile across the river. Fresno scrapers cleared away the brush, and the old *Advance* gold dredge was resurrected to excavate for the foundation. Then concrete retaining walls were built, running the length of the dam, and were filled with rock and earth. The Mexican-Colorado Navigation Company got the supply contract for the site, and finding that the *St. Vallier* by itself was inadequate for the task, they leased the *Cochan* from Captain Mellon and put both boats to work full-time.[23]

White and Company had contracted to complete the dam within two years, but, hindered by high water and inadequate supply, the work progressed much more slowly than expected. After a year and a half with the dam only one-third completed, the company gave up the contract and the U.S. Reclamation Service had to take over construction. Canceling the steamer contract, the government engineers had a railroad spur run from the Southern Pacific line to the dam. This greatly hastened construction and on 27 March 1909 the first dam across the Colorado River was finally completed.[24]

By then the steamboat business was already dead. The Mexican-Colorado company had folded after losing both the dam supply business and their boat, the *St. Vallier,* which had sunk in May 1907. Captain Mellon salvaged the boat and sold it to E. F. Sanguinetti, a Yuma merchant, for $750. Just two weeks before the completion of the dam, the *St. Vallier* broke loose from her moorings and sank again. This time she was declared a total loss and was dynamited to clear the channel.[25]

Mellon and his partners had also given up on the steamer business, selling the *Cochan* to the U.S. Reclamation Service on 2 March 1909, less than a month before the dam was completed. After forty-six years on the river, Mellon bid a sad, terse farewell to his friends and left for San Diego, choosing not to wait for completion of the dam that killed the river as he had known it.[26]

While the Southern Pacific railroad was struggling to close the canal intake the U.S. Reclamation Service was engaged in an even more ambitious effort to permanently dam the Colorado and irrigate the bottomlands. The old *Advance* gold dredge was resurrected and refitted in 1905 to work on Laguna Dam construction.

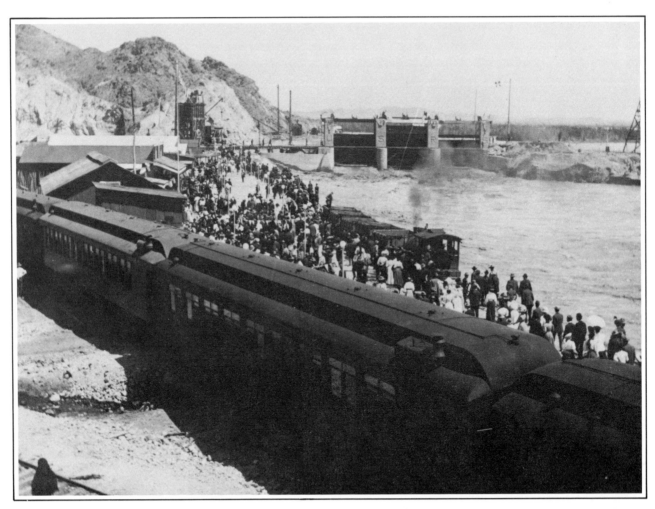

On 31 March 1909 trainloads of celebrants came to the U.S. Reclamation Service's ceremonies marking the completion of the Laguna Dam—*and* the end of the steamboat business on the Colorado River.

The California Development Company's receivers soon sold the *Searchlight* to the U.S. Reclamation Service. Having little use for either boat the government dismantled the *Cochan* in the spring of 1910. the *Searchlight,* with her upper deck screened in against mosquitoes, was maintained for a time to aid in levee construction along the river. The last steamboat on the Colorado, she was reported "lost on the river" on 3 October 1916.[27]

Thus after nearly two-thirds of a century the era of steam navigation on the Colorado finally came to a close. The last steamboat was gone, and there were few of her pioneers left to mourn her passing. George Johnson, who had built the Colorado steamboat business into the largest enterprise in Arizona, had died 27 November 1903 while the river trade was still experiencing its last revival. Most others prominent in the business were also dead. Even Steamboat Adams who had returned to the East to try to win a pension from the Congress had gone to his reward. Only Isaac Polhamus and Jack Mellon outlived the steamboats they served, and they had to live a lot longer than most to do so. Polhamus died 16 January 1922 at the age of ninety-four and Mellon on 17 December 1924 at eighty-three.[28]

The *Gila* and two barges loading at the Yuma landing just above the Southern Pacific railroad bridge in 1879.

Appendixes

Appendix A: Steamboats on the Colorado River and Its Tributaries

Name	Type	Tons	Length (ft.)	Beam (ft.)	Launched	Disposition
Black Eagle	screw	. . .	40	6	Green River, Utah Jun. 1907	Exploded 1907
Charles H. Spencer	stern	. . .	92.5	25	Warm Creek, Ariz. Feb. 1912	Abandoned Spring 1912
Cliff Dweller	stern	. . .	70	20	Halverson's, Utah Nov. 1905[1]	To Salt Lake[2] Apr. 1907
Cochan	stern	234	135	31	Yuma, Ariz. Nov. 1899	Dismantled Spring 1910
Cocopah (I)	stern	. . .	140	29	Gridiron, Mex. Aug. 1859	Dismantled[3] 1867
Cocopah (II)	stern	231	147.5	28	Yuma, Ariz.? Mar. 1867	Dismantled 1881

Appendixes

Name	Type	Tons	Length (ft.)	Beam (ft.)	Launched	Disposition
Colorado (I)	stern	...	120	...	Estuary, Mex. Dec. 1855	Dismantled Apr. 1862
Colorado (II)	stern	179	145	29	Yuma, Ariz. May 1862	Dismantled Aug. 1882
Comet	stern	...	60	12	Green River, Wyo. July 1908	Abandoned 1908
Esmeralda	stern	...	93	20	San Francisco 1862[4]	Dismantled 1868
Explorer	stern	...	54	13	Robinson's, Mex. Dec. 1857	Engine Removed 1858[5]
General Jesup	side	...	104	17	Estuary, Mex. Jan. 1854	Dismantled 1859
General Rosales	screw	...	...	...	Yuma, Ariz. July 1878	To Guaymas Sept. 1878
Gila	stern	236	149	31	Port Isabel, Mex. Jan. 1873	Rebuilt as *Cochan* 1899
Major Powell	screw	...	35	8	Green River, Utah Aug. 1891	Dismantled 1894
Mohave (I)	stern	193	135	28	Estuary, Mex. May 1864	Dismantled 1875
Mohave (II)	stern	188	149.5	31.5	Port Isabel, Mex. Feb. 1876	Dismantled Jan. 1900

Name	Type	Tons	Length	Beam	Launched	Disposition
Nina Tilden	stern	120	97	22	San Francisco July 1864[6]	Wrecked Sept. 1874
Retta	stern	...	36	6	Yuma, Ariz. 1900	Sunk Feb. 1905
St. Vallier	stern	92	74	17	Needles, Calif. Early 1899	Sunk Mar. 1909
San Jorge	screw	...	38	9	Yuma, Ariz. June 1901	To Gulf July 1901
Searchlight	stern	98	91	18	Needles, Calif. Dec. 1902	"Lost" Oct. 1916
Uncle Sam	side	40	65	16	Estuary, Mex. Nov. 1852	Sunk May 1853
Undine	stern	...	60	10	Green River, Utah Nov. 1901	Wrecked May 1902

[1] Rebuilt from gasoline-powered screw *City of Moab*.
[2] Renamed *Vista*.
[3] Made into a boardinghouse at Port Isabel.
[4] Reached the Colorado River in March 1864.
[5] Used as a barge until she sank in 1864.
[6] Reached the Colorado River in August 1864.

Many other early craft besides steamboats plied the Colorado and its tributaries including the gasoline boats: *Aztec* (I and II), *Baby Black Eagle, Betsy May, City of Moab, Colorado, Electric, Electric Spark, Hercules, Ida B., Iola* (I and II), *Katy Lloyd, Little Dick, Lucy B., Marguerite, Mohave* (III), *Mullins, Navajo, Paddy Ross, Sunbeam, Teddy R., Violet Louise, Water Pearl,* and *Wilmont;* the barges: *Arizona, Barge No. 1, No. 2, No. 3,* and *No. 4, Black Crook, Colorado, El Dorado, Enterprise, Pumpkin Seed, Silas J. Lewis, Veagas, Victoria, White Fawn* and *Yuma;* the sloop: *Sou'wester;* and the dredges: *Advance, Alpha, Beta, Delta, Hoskaninni* and *North Dakota.*

Appendix B: Chronological List of Steamboat Operators on the Colorado and Green Rivers

Operator		Boats
James Turnbull	1852–53	*Uncle Sam*
George A. Johnson & Co. (G. A. Johnson, B. M. Hartshorne and A. H. Wilcox)	1854–69	*Cocopah, Colorado, General Jesup, Mohave, Nina Tilden*
U.S. Army Topographical Engineers (Lt. J. C. Ives Expedition)	1857–58	*Explorer*
Gila Mining & Transportation Co.	1859	Unnamed steamer*
Union Line (T. E. Trueworthy et al.)	1864–65	*Esmeralda*
Philadelphia Silver & Copper Mining Co. (A. F. Tilden, manager)	1864–65	*Nina Tilden*
Pacific & Colorado Steam Navigation Co. (J. W. Stow, Pres., K. C. Eldredge, Sec.)	1865–66	*Esmeralda* *Nina Tilden*
Arizona Navigation Co. (Creditors of Pac. & Colo. Steam Nav. Co.)	1866–67	*Esmeralda* *Nina Tilden*

Operator		Boats
Colorado Steam Navigation Co. (G. A. Johnson, B. M. Hartshorne, A. H. Wilcox, E. Norton and R. D. Chandler, 1869–77; Western Development Co., 1877–86; I. Polhamus and J. A. Mellon, 1886–1904; J. A. Mellon, J. Gandolfo and J. Thornton, 1904–09)	1869–1909	*Cochan* *Cocopah* *Colorado* *Gila* *Mohave* *Nina Tilden*
Gulf of California Steamship Co. (T. H. Blythe & G. Andrade)	1878	*General Rosales*
Stacy Bros. (E. E. and O. T. Stacy)	1891–95	*Aztec* (gas) *Electric* (gas) *Electric Spark* (gas)
Colorado River & Gulf Transportation Co. (W. G. Purdy et al.)	1892	Unnamed steam launch
Santa Ana Mining Co.	1899–1900	*St. Vallier*
Mexican-Colorado Navigation Co. (A. B. Smith, W. S. Twogood and E. E. Busby)	1901–07	*Retta* *St. Vallier* *San Jorge*
Lamar Bros. (C. P. and L. F. Lamar)	1901–06	*Aztec* (gas)
Colorado River Transportation Co. (F. L. Hawley and F. L. Forrester)	1902–05	*Searchlight*
California Development Co.	1905–09	*Searchlight*

Appendixes

Operator		Boats
C. S. Hall	1906–08	*Iola* (gas)
Needles Navigation Co. (H. J. Caldwell and D. W. Tungate)	1907–08	*Hercules* (gas)
U.S. Reclamation Service	1909–16	*Cochan* *Searchlight*

Canyon Country

Operator		Boats
Green Grand and Colorado River Navigation Co. (B. S. Ross et al.)	1890–93	*Major Powell*
Frank H. Summeril	1901–02	*Undine*
Green-Grand River & Moab Navigation Co. (J. J. Lumsden and H. K. Clover)	1904–07	*City of Moab* (gas) *Cliff Dweller*
Harry T. Yokey	1907	*Black Eagle*
Green River Navigation Co. (M. N. Larsen et al.)	1908	*Comet*
Charles H. Spencer	1911–12	*Charles H. Spencer*

*Lost at the mouth of the Colorado before she was put in operation.

Appendix C: Table of Distances Along the Lower Colorado River

(Distances vary as the course of the river changes with time.)

	Miles from Yuma		Miles from Yuma		Miles from Yuma
Port Isabel, Sonora	157	Hanlon's Ferry, California	7	Aciquia, California	32
Mouth of the River	150	El Rio, California	5	Sakey, California	32
Robinson's Landing, Baja California	140	Jaeger's Ferry, California	1	Castle Dome, Arizona	35
Hardy's Colorado	127	Colorado City, Arizona	1	Welcome Ranch, California	40
Head of Tidewater	103	Fort Yuma, California	0	Yuma Arroya, Arizona	43
Heintzelman's Point	103	Yuma (Arizona City), Arizona	0	Hinton Island	43
Port Famine, Sonora	100	José Ranch, California	7	Hardscrabble Canyon, California	44
Lerdo Landing, Sonora	97	Willow Camp, Arizona	12	Eureka, Arizona	45
Gridiron, Sonora	83	Boat Knee Bend, Arizona	12	Williamsport, Arizona	47
Ogden's Landing, Sonora	55	Pot Holes, California	18	Chimney Peak, California	48
Hualapai Smith's, Sonora	45	Laguna, Arizona	20	Picacho, California	48
Pedrick's, Arizona	31	Buena Vista, California	28	Reliance Landing, California	48
Algodones, Baja California	8	Stevenson's Island	28	The Barriers	50

	Miles from Yuma
Johnson's Landing, Arizona	50
Duff's Ferry, Arizona	52
Norton's Landing, Arizona	52
Pacific City, Arizona	52
Carissa Arroya, California	55
Red Rock Gate	56
Sulphur Bend	58
Crawfords, California	58
Sacaton, Arizona	62
Light House Rock	63
Clip Landing, Arizona	70
Rood's Ranch, Arizona	71
Camp California, California	72
Redondo's, California	75
Camp Leon, California	77
Camp Gaston, California	80
Taylor's Camp, California	83
Carroll's Creek, California	85
Myers' Landing, California	94
Alvarez Ranch, Arizona	96
Drift Desert, Arizona	102
Swallows' Nest Bend	110
G. W. Brown's, California	112
Potato Point, Arizona	114
Arctic Landing, Arizona	116
Brown's Upper Ranch, Arizona	116
Alamo Ranch, California	119
Swan Lagoon, Arizona	120
Mineral City, Arizona	123
Bradshaw's Ferry, California	123
Ehrenberg, Arizona	125
Olive City, Arizona	127
La Paz, Arizona	131
Blythe's, California	137
310 Landing, California	137
Black Point, California	140
Quien Sabe Ranch, Ariz. & Calif.	150
Quien Sabe, California	157
New Slough, California	175
Riverside Mountain, California	180
Rattlesnake Point, Arizona	183
Riverside Ranch, California	188
Mesa Bend	190
Osborne's Ranch, California	196
Colorado Indian Reservation, Ariz.	200
Camp Colorado, Arizona	200
Parker's Landing, Arizona	200
Dent's Landing, Arizona	200
Beaver Island	202
Parker, Arizona	203
Planet Wash, Arizona	205
Paymaster's Bend, California	206
Empire Flat, Arizona	210
Iratata Flat, Arizona	216
Aubrey City, Arizona	220
Pedrigal, Arizona	222
Miller's Landing, Arizona	228
Murray's Mine, California	229
Boat Rock	230
Kelley's, California	234
Chimehuevis, California	240
Chims Valley, California	240
Liverpool Landing, California	242

	Miles from Yuma
Polhamus' Ranch, Arizona	245
Grand Turn	257
The Needles, Arizona	257
Mellen, Arizona	267
Beal, California	270
Powell, Arizona	271
Hope Landing, California	272
Poverty Bar, California	275
Pigeon Ranch, California	277
Needles, Arizona	281
Needles, California	282
Hood's Landing, California	282
Halfway Bend, California	282
Peas Ranch, Arizona	285
Gravel Point, California	288
Nevada Point, California	296
Iretaba City, Arizona	298
Fort Mohave, Arizona	300
Joaquin's Ranch, Arizona	301
Beal's Crossing, Nevada	303
Mohave City, Arizona	305
Hardyville, Arizona	310
Camp Alexander, Arizona	312
Polhamus Landing, Arizona	314
Cottonwood Island	339
Quartette Landing, Nevada	342
Murphyville, Arizona	353
Eldorado Canyon, Nevada	365
Explorer's Rock	372
Roaring Rapids	380
Vegas Wash, Nevada	402
Callville, Nevada	408
Stone's Ferry, Nevada	437
Virgin River, Nevada	440
Freemansburg, Nevada	440
Rioville, Nevada	440

The *St. Vallier* tied up at Needles awaiting cargo about 1901.

Notes to the Chapters

Opening the River

1. The principal studies of the opening of steam navigation on the Colorado River are Francis H. Leavitt, "Steam Navigation on the Colorado River," *California Historical Society Quarterly* 22 (1943):1–25, 151–74 and Arthur Woodward, *Feud on the Colorado* (Los Angeles: Westernlore Press, 1955).

2. *Arizona Sentinel (Yuma)*, 11 Aug. 1877; *San Diego Union*, 18 July 1885.

3. George A. Johnson, "Life of Captain George A. Johnson," typescript, California State Library, Sacramento, pp. 1–6; George A. Johnson, "The Steamer General Jesup," *Quarterly of the Society of California Pioneers* 9 (1932):109.

4. *San Diego Union*, 18 July 1885.

5. Thomas W. Sweeny, *Journal of Lt. Thomas W. Sweeny, 1849–1853*, Arthur Woodward, ed. (Los Angeles: Westernlore Press, 1956), pp. 49–51; Johnson, "Life," pp. 6–7; Benjamin Hayes, "Emigrant Notes," pt. 4, Bancroft Library, University of California, Berkeley, p. 745.

6. U.S., Congress, House, *House Executive Document* 90, "Report upon the Colorado River of the West, Explored in 1857 and 1858," 36th Cong., 1st sess., prepared by Joseph C. Ives (Washington, D.C.: Government Printing Office, 1861), p. 43; U.S., Congress, House, *House Executive Document* 135, vol. 1, "Report of the United States and Mexican Boundary Survey," 34th Cong., 1st sess., prepared by William H. Emory (Washington, D.C.: Government Printing Office, 1857), p. 106; Bancroft Scraps, Arizona Miscellany, Bancroft Library, University of California, Berkeley, p. 335.

7. *San Francisco Herald*, 22 Oct. 1851; *Arizona Sentinel* (Yuma), 28 July 1877; U.S., Congress, House, *House Executive Document* 81, "Report of a Reconnaissance of the Gulf of California and the Colorado River," 32d Cong., 1st sess., prepared by George H. Derby (Washington, D.C.: Government Printing Office, 1852), pp. 2–3.

8. *House Executive Document* 81, 32 Cong., 1 sess., pp. 13, 16–17; Benjamin Hayes, Scrapbook, "Colorado River," Bancroft Library, University of California, Berkeley, p. 120.

9. *House Executive Document* 81, 32 Cong., 1 sess., pp. 18–19; *House Executive Document* 90, 36 Cong., 1 sess., p. 28.

10. *House Executive Document* 81, 32 Cong., 1 sess., p. 20; *San Francisco Daily Herald*, 22 Oct. 1851.

11. Sweeny, *Journal*, pp. 51–52, 117, 137.

12. Johnson, "Life," pp. 7–9; Johnson, "General Jesup," pp. 109–10; *Los Angeles Star*, 30 Aug. 1851; *Alta California* (San Francisco), 8 Apr., 10 May 1852; *Arizona Sentinel* (Yuma), 18 May 1878.

13. *San Diego Herald*, 28 June 1852, quoted in *Alta California* (San Francisco), 11 July 1852; *Arizona Sentinel* (Yuma), 25 Aug. 1877, 18 May 1878.

14. *Arizona Sentinel* (Yuma), 18 May 1878; *San Francisco Herald*, 31 Dec. 1852; Sweeny, *Journal*, p. 187.

15. Sweeny, *Journal,* pp. 187–91; *Los Angeles Star,* 19 Mar. 1853.
16. *Los Angeles Star,* 25 Dec. 1852; *San Francisco Herald,* 31 Dec. 1852; Sweeny, *Journal,* p. 187.
17. Sweeny, *Journal,* p. 188.
18. *Ibid.,* pp. 189–90.
19. Sweeny, *Journal,* p. 207; *San Diego Herald,* 9 July 1853; *San Francisco Herald,* 11 June 1853.
20. *San Francisco Herald,* 11 June 1853; *San Diego Herald,* 24 Sept. 1853; *Arizona Sentinel* (Yuma), 25 Aug. 1877.
21. *Arizona Sentinel* (Yuma), 25 May 1878; *Alta California* (San Francisco), 9 Apr. 1854.
22. Albert Jacob Johnson, letter to his mother, Camp Yuma, 12 Dec. 1856, Johnson Papers, Arizona Historical Society Library, Tucson; *Alta California* (San Francisco), 12, 19 Sept., 14 Dec. 1854; *Arizona Sentinel* (Yuma), 25 May 1878.
23. Charles G. Johnson, *History of the Territory of Arizona and the Great Colorado of the Pacific* (San Francisco: Vincent Ryan and Co., 1868), pp. 8–14.
24. Charles Johnson, *History of the Territory of Arizona,* pp. 5–6; *House Executive Document* 90, 36 Cong., 1 sess., p. 27; *Arizona Sentinel* (Yuma), 1 Aug. 1874.
25. *Arizona Sentinel* (Yuma), 25 Aug. 1877; *Alta California* (San Francisco), 27 July 1859; Hayes, Scrapbook, "Colorado River," pp. 185, 192; *War of the Rebellion; A Compilation of the Official Records of the Union and Confederate Armies,* ser. 1, vol. 50, pt. 1 (Washington, D.C.: Government Printing Office, 1897), pp. 815–17.
26. Albert Johnson, letter to his mother, 12 Dec. 1856; Sweeny, *Journal,* p. 213; Hayes, Scrapbook, "Colorado River," p. 185; Bancroft Scraps, Southern California, San Diego, Bancroft Library, University of California, Berkeley, p. 270; Sylvester Mowry, *Arizona and Sonora* (New York: Harper Bros., 1864), p. 82.
27. *Sacramento Daily Union,* 30 Aug. 1855; *Alta California* (San Francisco), 27 Dec. 1855; *San Diego Herald,* 22 Dec. 1855; *Arizona Sentinel* (Yuma), 25 Aug. 1877; *House Executive Document* 81, 32 Cong., 1 sess., p. 20; *Santa Fe Gazette,* 10 May 1856.
28. *Los Angeles Star,* 1 Jan. 1856; Hayes, Scrapbooks, "Navigation of the Colorado," p. 181.
29. Johnson, "General Jesup", p. 110–11; *House Executive Document* 90, 36 Cong., 1 sess., p. 6; Hayes, Scrapbooks, "Navigation of the Colorado," p. 190.
30. *House Executive Document* 90, 36 Cong., 1 sess., pp. 21–37.
31. *House Executive Document* 90, 36 Cong., 1 sess., p. 38; Johnson, "General Jesup," p. 111; Woodward, *Feud,* pp. 97–104; *Arizona Sentinel* (Yuma), 1 June 1878; *San Francisco Herald,* 14 Mar. 1858.
32. *San Francisco Herald,* 14 Mar. 1858.
33. *San Francisco Herald,* 14 Mar. 1858; Hayes, "Emigrant Notes," pp. 744–45, 754–55.
34. U.S., Congress, House, *House Executive Document* 124, "Report of Edward Fitzgerald Beale to the Secretary of War Concerning the Wagon Road from Fort Defiance to the Colorado River," 35th Cong., 1st sess. (Washington, D.C.: Government Printing Office, 1855), pp. 76–77.
35. *House Executive Document* 90, 36 Cong., 1 sess., p. 46.
36. *Ibid.,* pp. 47–56.
37. *House Executive Document* 90, 36 Cong., 1 sess., pp. 44, 56; *San Francisco Herald,* 14 Mar. 1858.
38. *San Francisco Herald,* 11, 14 Mar. 1858.
39. *House Executive Document* 90, 36 Cong., 1 sess., pp. 81–82, 87.
40. *Ibid.,* pp. 83–87.
41. *Ibid.,* pp. 88–90.
42. *House Executive Document* 90, 36 Cong., 1 sess., p. 91; *San Diego Herald,* 31 July 1858; *Alta California* (San Francisco), 27 July 1859; *Arizona Sentinel* (Yuma), 28 Sept. 1878; Hayes, "Emigrant Notes," p. 744; Godfrey Sykes, *The Colorado Delta* (New York: American Geographical Society, 1937), pp. 90–92.
43. *House Executive Document* 124, 35 Cong., 1 sess., p. 76.
44. *House Executive Document* 124, 35 Cong., 1 sess., p. 2; *Los Angeles Star,* 13 Nov. 1858; *Alta California* (San Francisco), 7 May 1859; U.S., Congress, Senate, *Senate Executive Document* 2, pt. 2, "Report of the Secretary of War, Affairs in Dept. of California," 36th Cong., 1st sess. (Washington, D.C.: Government Printing Office, 1861), pp. 387, 410.
45. *Senate Executive Document* 2, pt. 2, 36 Cong., 1 sess., pp. 389–92.
46. *Los Angeles Star,* 26 Mar. 1859; George A. Johnson, letter to his father, Fort Yuma, 27 Mar. 1859, Johnson Papers, Arizona Historical Society Library, Tucson; *Arizona Sentinel* (Yuma), 25 May 1878; *Alta California* (San Francisco), 6, 7 May 1859.

47. *Alta California* (San Francisco), 6, 7 May 1859; George A. Johnson, letter to his father, 27 Mar. 1859; *Senate Executive Document 2*, pt. 2, 36 Cong., 1 sess., p. 410.

48. *Senate Executive Document 2*, pt. 2, 36 Cong., 1 sess., pp. 416–17, 419–22.

49. *Senate Executive Document 2*, pt. 2, 36 Cong., 1 sess., pp. 416–17; *Los Angeles Star,* 22 Oct., 31 Dec. 1859.

50. George A. Johnson, letter to his father, 27 Mar. 1859; *Alta California* (San Francisco), 22, 26 May 1859; *Arizona Miner* (Prescott), 25 May 1864; *Arizona Sentinel* (Yuma), 28 Sept. 1878; *Tucson Citizen,* 17 Jan. 1922.

The Arizona Fleet

1. The first history of Colorado River steamboating covering this period was "The Arizona Fleet," *Arizona Sentinel* (Yuma), 28 Sept. 1878.

2. Richard J. Hinton, *The Handbook to Arizona* (San Francisco: Payot, Upham & Co., 1878), p. 154; *San Diego Herald,* 13 Nov. 1858; George A. Johnson, letter to his father, George W. Johnson, Fort Yuma, 27 Mar. 1859, Johnson Papers, Arizona Historical Society Library, Tucson; Benjamin Hayes Scrapbooks, "Colorado River," Bancroft Library, University of California, Berkeley, p. 193.

3. *Alta California* (San Francisco), 4, 19, 22, 26 Jan. 1859; *Los Angeles Star,* 4 Dec. 1858, 8 Jan. 1859, 24 June 1860; J. Ross Browne, *Adventures in Apache Country* (New York: Harper Bros., 1869), p. 77.

4. *Alta California* (San Francisco), 12 Feb. 1859.

5. G. O. Haller to W. W. Mackall, 30 Apr. 1861, U.S. National Archives, Record Group 393, Fort Mohave Letters Sent (copy courtesy of Dennis G. Casebier); *Los Angeles Star,* 1 Nov. 1861, 29 Mar., 1 Nov. 1862; *Mining & Scientific Press* (San Francisco), 21 Oct. 1865; *Alta California* (San Francisco), 22 Sept. 1871; Bancroft Scraps, "Arizona Miscellany," Bancroft Library, University of California, Berkeley, pp. 291–92; Rossiter W. Raymond, *Mineral Statistics West of the Rocky Mountains* (Washington, D.C.: Government Printing Office, 1870) p. 266.

6. *Mining and Scientific Press* (San Francisco), 16 Feb. 1867; Bancroft Scraps, Arizona Miscellany, pp. 291–92.

7. Bancroft Scraps, Arizona Miscellany, pp. 288, 292, 323, 328.

8. Bancroft Scraps, Arizona Miscellany, pp. 288, 295, 321; *Mining & Scientific Press* (San Francisco), 2 Apr. 1864, 20 May 1865; *Arizona Miner* (Prescott), 14 Sept. 1867.

9. Bancroft Scraps, "Arizona Miscellany," pp. 271, 284–85.

10. Bancroft Scraps, "Arizona Miscellany," pp. 271, 279, 284–86, 291, 326.

11. *Ibid.,* pp. 276, 279, 291–92, 296, 325–26, 500.

12. Bancroft Scraps, "Arizona Miscellany," pp. 285, 292–93, 305, 314, 325; *Arizona Sentinel* (Yuma), 1 Sept. 1877.

13. Bancroft Scraps, "Arizona Miscellany," pp. 279, 286, 302, 310, 326.

14. Bancroft Scraps, "Arizona Miscellany," pp. 293, 310–11; *Los Angeles Southern News,* 25, 28 Nov. 1863; *Mining & Scientific Press* (San Francisco), 2 Apr. 1864.

15. *Los Angeles Star,* 28 Mar., 9 May, 19 Dec. 1863; *Mining & Scientific Press* (San Francisco), 9 Apr., 21 Oct. 1865; Bancroft Scraps, "Arizona Miscellany," pp. 296–97, 302, 317–19.

16. Bancroft Scraps, "Arizona Miscellany," pp. 207, 317; George A. Johnson to Mr. Morse, Fort Yuma, 10 Apr. 1865, Johnson Papers, Arizona Historical Society Library, Tucson.

17. Bancroft Scraps, "Arizona Miscellany," pp. 294–95, 316, 326–27.

18. Bancroft Scraps, "Arizona Miscellany," pp. 86, 272, 295, 302, 319, 321; *Los Angeles Southern News,* 14 Nov. 1863.

19. Bancroft Scraps, "Arizona Miscellany," pp. 297–98, 301, 317, 333.

20. George A. Johnson letter, 10 Apr. 1865; Albert Johnson letter, 3 Feb. 1870, Johnson Papers, Arizona Historical Society Library, Tucson; *Sacramento Union,* 29 Sept. 1862; *Tucson Arizonian,* 14 Mar. 1869.

21. Bancroft Scraps, "Arizona Miscellany," p. 186; *San Diego Union,* 9 July 1862; *Arizona Miner* (Prescott), 25 May 1864.

22. *Arizona Miner* (Prescott), 25 May 1864.

23. Bancroft Scraps, "Arizona Miscellany," p. 313.

24. *Ibid.,* p. 326.

25. *Los Angeles Southern News,* 16, 18 Dec. 1863; *Arizona Miner* (Prescott), 6 Apr. 1864; *Arizona Sentinel* (Yuma), 15 May 1875; U.S. Congress, Senate, *Senate Document 13,* "Federal Census—Territory of New Mexico and Territory of Arizona" 89th Cong., 1st sess., (Washington, D.C.: Government Printing Office, 1965), p. 105.

26. *Arizona Magazine,* Aug. 1893, pp. 59–68; Hayes Scrapbook, "Colorado River," p. 171; *Arizona Sentinel* (Yuma), 28 Sept. 1878.

27. *Arizona Sentinel* (Yuma), 28 Sept. 1878; *Arizona Miner* (Prescott), 25 Feb. 1864, 8 Aug. 1866; Hayes, Scrapbooks, "Colorado River," p. 194.

28. Hayes, Scrapbooks, "Colorado River," pp. 197–98; *California Express* (San Francisco), 23 July 1864; *San Francisco Democratic Press,* 1 Aug. 1864; *Alta California* (San Francisco), 6 Aug. 1864; *Los Angeles News,* 18 Oct. 1864.

29. *Arizona Miner* (Prescott), 25 May 1864, 4 Jan. 1873; *Alta California* (San Francisco), 21 June 1866; Bancroft Scraps, "Arizona Miscellany," p. 438.

30. Bancroft Scraps, "Arizona Miscellany," p. 493; *Alta California* (San Francisco), 4 Aug. 1864; U.S. Congress, House, *House Miscellaneous Document* 12, "The Exploration of the Colorado River and its Tributaries," prepared by Samuel Adams, 41st Cong., 3d sess., (Washington, D.C.: Government Printing Office, 1871); U.S. Congress, House, *House Executive Document* 166, "Freight to Salt Lake City by the Colorado River" 42d Cong., 2d sess. (Washington, D.C.: Government Printing Office, 1872), p. 6.

31. *Salt Lake Daily Telegraph, 3 Jan. 1865; Deseret News* (Salt Lake), 29 Mar. 1865; Hayes, Scrapbooks, "Colorado River," p. 202; Bancroft Scraps, "Arizona Miscellany," p. 437; *United States v. Utah,* Complainant's Exhibit 624, Church Historian's Office, Salt Lake City.

32. *Alta California* (San Francisco), 7 Jan., 25 Apr. 1865; *Deseret News* (Salt Lake), 29 Mar. 1865; Hayes, Scrapbooks, "Colorado River," p. 202; *House Executive Document* 166, 41 Cong., 3 sess; *House Executive Document* 12, 42 Cong., 2 sess.

33. *House Executive Document* 12, 42 Cong., 2 sess., p. 2; *Mining & Scientific Press* (San Francisco), 28 Oct. 1865; *Deseret News* (Salt Lake), 11 July 1866; Hayes, Scrapbooks, "Colorado River," p. 201; Bancroft Scraps, "Arizona Miscellany," pp. 437–38.

34. Bancroft Scraps, "Arizona Miscellany," pp. 170, 338, 343–44, 438.

35. *House Executive Document* 166, 41 Cong., 3 sess., p. 2; Hayes, Scrapbooks, "Colorado River," p. 178, 201.

36. Hayes, Scrapbooks, "Colorado River," pp. 178–97, 204–5, 442; *San Francisco Evening Bulletin,* 1 Oct. 1867.

37. *House Executive Document* 166, 41 Cong., 3 sess., pp. 6–7; *House Executive Document* 12, 42 Cong., 2 sess., p. 4.

38. *House Executive Document* 12, 42 Cong., 2 sess., p. 4; *Deseret News* (Salt Lake), 16 June 1869; Hayes, Scrapbooks, "Colorado River," pp. 187, 193.

39. Frances Fleming, "The History of Steam Transportation on the Colorado River," M.A. Thesis, Arizona State College, 1950, pp. 49–51.

40. *Senate Document* 13, 89 Cong., 1 sess., p. 233–46; *Arizona Gazette* (Phoenix), 10 Apr. 1895; Hayes, Scrapbooks, "Colorado River," p. 208; Francis Berton, *A Voyage on the Colorado, 1878,* ed. and trans. by Charles N. Rudkin (Los Angeles: Glen Dawson, 1953), p. 54.

41. *House Executive Document* 166, 41 Cong., 3 sess., p. 3; *House Executive Document* 12, 42 Cong., 2 sess., p. 11; Charles G. Johnson, *History of the Territory of Arizona and the Great Colorado of the Pacific* (San Francisco: Vincent Ryan & Co., 1868), pp. 6–7; *Arizona Sentinel* (Yuma), 5 July 1873.

42. *Arizona Sentinel* (Yuma), 5 July 1873, 1 Aug. 1874.

43. *Arizona Sentinel* (Yuma), 28 Sept. 1878.

44. *Arizona Sentinel* (Yuma), 8 Feb. 1873, 4 Mar. 1876; *Merchant Vessels of the United States* (Washington, D.C.: Government Printing Office, 1895), pp. 250, 279; Martha Summerhayes, *Vanished Arizona* (Philadelphia: J. B. Lippincott Co., 1908), p. 309.

45. *Arizona Sentinel* (Yuma), 20 Jan. 1877.

46. *Arizona Sentinel* (Yuma), 4 Mar. 1876.

47. Bancroft Scraps, "Arizona Miscellany," p. 117.

48. *Ibid.* pp. 11, 49.

49. Charles Johnson, *History,* pp. 13–15; *Arizona Sentinel* (Yuma), 8 Feb. 1873, 21 Feb. 1874.

50. *Arizona Sentinel* (Yuma), 10 Apr. 1875; Berton, *Voyage,* p. 58.

51. Berton, *Voyage,* p. 26; Bancroft Scraps, Arizona Miscellany, p. 124.

52. *Arizona Sentinel* (Yuma), 29 Dec. 1877; *Crofutt's New Overland Tourist and Pacific Coast Guide,* v. 1 (Chicago: Overland Publishing Co., 1878–79), p. 255; descriptions of steamer trips on the river are found in Berton, *Voyage,* pp. 26ff.; Summerhayes, *Vanished Arizona,* pp. 22ff.; *Arizona Sentinel* (Yuma), 30 Aug. 1873, 20 June 1874, 12 Feb. 1876; and Bancroft Scraps, "Arizona Miscellany," pp. 125–27.

53. Bancroft Scraps, "Arizona Miscellany," pp. 126, 372, 375–76; *Senate Document* 12, 89 Cong., 1 sess., p. 96; *Arizona Sentinel* (Yuma), 20 June 1874.

54. Berton, *Voyage,* pp. 39–40; Bancroft Scraps, "Arizona Miscellany," p. 126.

55. Berton, *Voyage,* p. 41; *Arizona Miner* (Prescott), 7 May 1870.

56. *Arizona Sentinel* (Yuma), 12 Feb. 1876; Summerhayes, *Vanished Arizona*, p. 39; Berton, *Voyage*, pp. 34, 44.

57. Berton, *Voyage*, p. 44.

58. Berton, *Voyage*, pp. 44–45; Summerhayes, *Vanished Arizona*, p. 40; *Arizona Sentinel* (Yuma), 12 Feb. 1876; U.S. Congress, House, *House Executive Document* 1, pt. 2, "Report of Chief of Engineers, Appendix JJ," 46th Cong., 2d sess. (Washington, D.C.: Government Printing Office, 1879).

59. *House Executive Document* 1, pt. 2, 46 Cong., 2 sess., p. 1779; Summerhayes, *Vanished Arizona*, pp. 38–40; Berton, *Voyage*, pp. 47, 51, 53, 56, 59, 86.

60. Summerhayes, *Vanished Arizona*, p. 41; *Arizona Sentinel* (Yuma), 5 May 1877, 9 Nov. 1878.

61. *Arizona Sentinel* (Yuma), 30 Aug. 1873, 4 May 1878; *Alta California* (San Francisco), 24 Jan. 1870; Bancroft Scraps, "Arizona Miscellany," pp. 126, 501.

62. Bancroft Scraps, "Arizona Miscellany," p. 126; Berton, *Voyage*, p. 76.

63. Berton, *Voyage*, pp. 78–79; Bancroft Scraps, "Arizona Miscellany," pp. 126, 157, 377, 391; *Arizona Sentinel* (Yuma), 30 Aug. 1873.

64. *Arizona Sentinel* (Yuma), 30 Aug. 1873.

65. Bancroft Scraps, "Arizona Miscellany," pp. 357–58, 363–65, 377; *Alta California* (San Francisco), 19 Jan. 1872.

66. *Arizona Sentinel* (Yuma), 27 Oct. 1877.

67. *Frontier Index* (Bear River City), 24 Mar., 28 Apr., 5, 26, 29 May, 2, 5, 9, 12, 16, 19 June, 7 July 1868; *The Piute Company of California and Nevada* (San Francisco: E. Bosqui and Co., 1870).

68. Hiram C. Hodge, *Arizona As It Is, Or the Coming Country* (New York: Hurd and Houghton, 1877), p. 210; *Arizona Miner* (Prescott), 4 Jan. 1873, 29 May 1874.

Through Progress of the Railroads

1. *Arizona Sentinel* (Yuma), 2, 16 June, 4, 11 Aug., 29 Sept., 6 Oct., 29 Dec. 1877; Bancroft Scraps, "Arizona Miscellany," pp. 407–86, Bancroft Library, University of California, Berkeley.

2. *Testimony Taken by the United States Pacific Railway Commission* (Washington, D.C.: Government Printing Office, 1887); *Arizona Sentinel* (Yuma), 17 Mar. 1877.

3. *Arizona Sentinel* (Yuma), 15 Dec. 1877, 12, 29 Jan., 13 Apr. 1878.

4. *Ibid.*, 2 Nov. 1877, 23 Mar., 13 Apr. 1878, 14, 28 Feb. 1880.

5. *Arizona Sentinel* (Yuma), 11, 18 May, 1 June, 13, 20 July, 3, 17 Aug., 7 Sept. 1878, 1 June 1881, 29 Apr., 26 Aug. 1882, 3 Mar. 1883.

6. *Ibid.*, 11, 18 May, 1 June, 13, 20 July 1878.

7. *Arizona Sentinel* (Yuma), 5 Aug. 1882. *Pacific Railway Commission*, p. 4; U.S. Congress, House, *House Document* 1, pt. 2, "Report of Chief of Engineers, Appendix JJ" 46th Cong., 2d sess., (Washington, D.C.: Government Printing Office, 1879), p. 1778.

8. John Bermingham to A. H. Wilcox, San Francisco, 3 May 1879, quoted in John Haskell Kemble, "To Arizona by Sea, 1850–1877" *Westerners Brandbook* (Los Angeles: L.A. Corral 1963), p. 152; *Arizona Sentinel* (Yuma), 21 June 1879.

9. *Arizona Sentinel* (Yuma), 21, 28 June, 5, 12, 19, 26 July 1879.

10. *Ibid.*, 19, 26 July 1879, 3 Mar. 1888.

11. *Arizona Sentinel* (Yuma), 18 Oct., 15 Nov. 1879, 23 July, 10 Sept. 1881, 14 Jan., 29 July 1882, 14 Apr. 1883; *Phoenix Saturday Review*, 13 Apr. 1895; U.S. Congress, House, *House Document* 67 "Preliminary Examination of the Colorado River, Nevada" 56th Cong., 2d sess., (Washington, D.C.: Government Printing Office, 1900), pp. 3, 6.

12. *Arizona Sentinel* (Yuma), 4, 25 June 1881.

13. Bancroft Scraps, "Arizona Miscellany," p. 126; *San Francisco Bulletin*, 28 Oct. 1878; *Arizona Sentinel* (Yuma), 28 Apr. 1877, 27 Apr., 1 June, 9, 16 Nov. 1878; 3 Apr. 1880, 30 Dec. 1882, 14 Apr. 1883, 20 June 1885.

14. *Arizona Sentinel* (Yuma), 25 Jan., 22 Feb., 15 Mar., 5 July, 13 Sept., 27 Dec. 1879, 6 Mar. 1880, 1, 8 Jan. 1881, 6 Jan., 14 Apr., 23 June 1883, 17 Mar. 1888.

15. *Arizona Sentinel* (Yuma), 31 Jan. 1880; *Arizona Miner* (Prescott), 4 Jan. 1873.

16. *Arizona Sentinel* (Yuma), 9, 23 June, 7 July 1883.

17. David F. Myrick, *Railroads of Nevada and Eastern California*, v. 2 (Berkeley: Howell-North, 1963), pp. 766–77.

18. *Arizona Sentinel* (Yuma), 2 May, 18 July 1885, 4 Sept. 1886; *Needles Eye*, 26 June 1906; L. A. Ingersoll, *Century Annals of San Bernardino County 1769 to 1904* (Los Angeles: L. A. Ingersoll, 1904) p. 776.

19. *Needles Eye,* 14 June 1891, 27 Mar. 1892, 12 May 1894; *Arizona Daily Gazette* (Phoenix), 27 June 1893; *Arizona Sentinel* (Yuma), 20 Oct., 24 Nov. 1888, 3 May 1890.

20. *Needles Eye,* 27 Mar., 22 May, 19 June 1892, 22 Apr. 1893; *Searchlight* (Searchlight, Nev.), 18 Aug. 1905; *Arizona Sentinel* (Yuma), 16, 23 July 1892, 9 Sept. 1893, 29 Sept. 1894.

21. *Arizona Sentinel* (Yuma), 23 Oct., 27 Nov., 11 Dec. 1901, 19 Mar., 21, 28 May 1902.

22. *Arizona Sentinel* (Yuma), 7 Mar. 1891, 23 Jan. 1893; *Arizona Republican* (Phoenix), 1 Jan. 1896.

23. *Needles Eye,* 4 Oct., 6 Dec. 1891; *Arizona Sentinel* (Yuma), 7, 21 Nov., 5, 12, 26 Dec. 1891, 9, 16, 30 Jan. 1892.

24. *Arizona Sentinel* (Yuma), 20, 27 Feb., 5, 19 Mar., 9 Apr., 7, 14, 21 May, 12, 26 Nov. 1892.

25. *Arizona Sentinel* (Yuma), 28 Jan., 11, 18 Feb., 4 Mar., 22 Apr., 2 Sept., 4 Nov. 1893, 19, 26 May, 29 Sept., 3, 10 Nov. 1894; *Merchant Vessels of the United States* (Washington, D.C.: Government Printing Office, 1895), p. 220.

26. *Arizona Sentinel* (Yuma), 22 Dec. 1894, 30 Mar. 1895, 22 May 1897, 22 Jan. 1898, 2 Dec. 1899; *Arizona Republican* (Phoenix), 1 Jan. 1896.

27. *Needles Eye,* 22 May, 12 June, 3, 24 July 1892.

28. *Needles Eye,* 3 June 1893; *Arizona Sentinel* (Yuma), 22 May 1897.

29. *Arizona Sentinel* (Yuma), 14 Jan., 2 Dec. 1899, 20 June 1900; *Merchant Vessels of the United States* (Washington, D.C.: Government Printing Office, 1900) p. 294.

30. *Merchant Vessels* (1900), p. 223; *Merchant Vessels of the United States* (Washington, D.C.: Government Printing Office, 1904), p. 195; *Merchant Vessels of the United States* (Washington, D.C.: Government Printing Office, 1907), pp. 364, 378; *Arizona Star* (Tucson), 23 Dec. 1908; *Needles Eye,* 13 Apr. 1907; *Los Angeles Mining Review,* 10 Feb. 1900; *Arizona Sentinel* (Yuma), 11 Nov. 1899, 13, 20, 31 Jan. 1900.

31. *Arizona Sentinel* (Yuma), 4 Nov. 1893, 25 Dec. 1897, 11, 18 Nov. 1899, 6 June 1900.

32. *Ibid.,* 20 Mar., 3, 10 Apr., 5, 26 June 1901.

33. *Ibid.,* 20 June 1900, 2 Jan., 19 June, 10 July 1901.

34. *Arizona Sentinel* (Yuma), 18 Dec. 1901, 15, 22, 29 Jan., 26 Feb., 21, 28 May, 4, 18 June 1902; *Needles Eye,* 8 Apr. 1893.

35. *Merchant Vessels* (1904), p. 297; *San Bernardino Weekly Times-Index,* 20 Mar., 3 Apr., 8, 15, 29 May, 26 June 1903; *Arizona Sentinel* (Yuma), 24 Dec. 1902; *Western Mining World* (Chicago), 6 Dec. 1902.

36. *Western Mining World* (Chicago), 31 Jan. 1903; *Los Angeles Mining Review,* 16 Feb. 1901; *Arizona Sentinel* (Yuma), 18 June 1902, 12 Oct. 1904, 22 Feb., 5 Apr. 1905, 29 Aug. 1906; *Needles Eye,* 20 Oct., 22 Dec. 1906, 24 June 1911.

37. *Yuma Sun,* 17 June 1910; *Arizona Republican* (Phoenix), 17 Jan. 1922.

38. *Arizona Sentinel* (Yuma), 17 Aug. 1904, 1, 8 Mar., 19 Apr., 23 Aug. 1905.

39. *Needles Eye,* 9, 23 June, 11 Aug., 27 Oct., 10 Nov. 1906, 26 Jan., 16 Mar., 15 June, 6 July, 24, 31 Aug. 1907, 20 Mar., 29 May 1909; *Arizona Sentinel* (Yuma), 29 July 1908.

40. *Los Angeles Mining Review,* 15 Dec. 1900.

41. *Los Angeles Mining Review,* 10 Feb., 2 June, 8 Sept. 1900; *Mining & Scientific Press* (San Francisco), 3 Feb., 12 May, 8 Sept. 1900; *Arizona Sentinel* (Yuma), 5 Aug., 2 Sept. 1899, 24 Jan., 28 Feb., 2, 30 May, 27 June, 5 Sept. 1900.

42. *Arizona Sentinel* (Yuma), 25 July, 22 Aug., 19 Sept. 1900, 29 May, 14 Aug. 1901; *Engineering & Mining Journal* (New York), 1 June 1901, 26 July, 20 Sept. 1902.

43. *Searchlight Bulletin,* 16 Aug. 1907, 17 Apr. 1908, 19 Mar., 11 June, 3 Dec. 1909.

44. *Ibid.,* 3 Dec. 1909, 7 Jan. 1910.

45. *Arizona Sentinel* (Yuma), 6 June 1900.

Steamboats in the Canyons

1. *Grand Valley Times* (Moab), 5, 19 Feb. 1904. A very brief account of steam navigation in the canyon country has been written by Frederick S. Dellenbaugh, *The Romance of the Colorado River* (New York: G. P. Putnam's Sons, 1909), p. 368.

2. *Grand Valley Times* (Moab), 6 Feb. 1903.

3. *Colorado Sun* (Denver), 3 July 1892; *Cheyenne Sun Annual,* 1 Nov. 1892; Dellenbaugh, *Romance,* p. 368.

4. *United States vs. Utah; Abstract in Narrative Form of the Testimony Taken before the Special Master,* (Washington, D.C.: Government Printing Office, 1931), pp. 561, 1242.

5. *Colorado Sun* (Denver), 3 July 1892.

6. *Denver Republican,* 4 June 1893; *United States vs. Utah,* pp. 555, 566.

7. *United States vs. Utah,* pp. 838, 1082.

8. *United States vs. Utah,* p. 566; *Denver Republican,* 4 June 1893.

9. *United States vs. Utah,* p. 570.

10. *Grand Valley Times* (Moab), 10 July, 20 Nov. 1896, 8 Jan., 12 Feb. 1897, 5 May 1899, 5 Feb. 1904, 2, 23 June, 11 Aug. 1905, 5 Jan. 1906.

11. *Grand Valley Times* (Moab), 13 Dec. 1901, *United States vs. Utah,* p. 1059.

12. *United States vs. Utah,* pp. 1059–62; U.S. Congress, House, *House Document* 953, "Grand and Green Rivers, Utah" 61st Cong., 2d sess., (Washington, D.C.: Government Printing Office, 1911), pp. 21, 23; *Grand Valley Times* (Moab), 13 Dec. 1901.

13. *Grand Valley Times* (Moab), 13, 20 Dec. 1901.

14. *Ibid.,* 7, 21 Feb., 14 Mar., 11 Apr. 1902.

15. *Ibid.,* 9, 16, 23 May 1902.

16. *Ibid.,* 30 May, 6 June 1902, 29 May, 19 June 1903, 12 Jan. 1906.

17. *Grand Valley Times* (Moab), 21 Oct., 4, 18 Nov. 1904, 27 Jan., 24 Feb., 10 Mar., 5 May, 11 Aug. 1905; *United States vs. Utah,* pp. 179–84, 196–99.

18. *United States vs. Utah,* pp. 180–82, 198–99; *Grand Valley Times* (Moab), 5 May 1905, 12 Jan. 1906.

19. *Grand Valley Times* (Moab), 26 May 1905; *United States vs. Utah,* pp. 182–83, 200–201.

20. *Grand Valley Times* (Moab), 14 July, 11 Aug, 3 Nov. 1905, 12 Jan., 11 May, 6 July 1906.

21. *Grand Valley Times* (Moab), 24 Aug. 1906; *United States vs. Utah,* pp. 201–3, 868–71.

22. *United States vs. Utah,* p. 871; *Grand Valley Times* (Moab), 6 July 1906, 19 Apr., 14, 21 June 1907.

23. *Grand Valley Times* (Moab), 9 Sept. 1904, 7 Apr. 1905, 13 Dec. 1907; *United States vs. Utah,* pp. 1372–77, 1391–92.

24. *United States vs. Utah,* pp. 1065–66; *Grand Valley Times* (Moab), 3 Nov. 1905, 22 Nov. 1907.

25. *Grand Valley Times* (Moab), 22 Nov. 1907, 7 May, 1 Oct. 1909, 1 Sept. 1911; *United States vs. Utah,* pp. 1194–99, 1377–81, 1392–93, 1396–97.

26. *United States vs. Utah,* pp. 307–27, 1331–39; *Grand Valley Times* (Moab), 18, 25 June, 6, 13 Aug., 24 Sept. 1909; Barbara Baldwin Ekker, "Freighting on the Colorado River," *Utah Historical Quarterly* 32, (1964): 122–29.

27. *Grand Valley Times* (Moab), 24 Apr. 1908; *Green River Star,* 20 Mar., 10 Apr., 12 June 1908.

28. *Green River Star,* 10, 17 July 1908; *United States vs. Utah,* pp. 1570–71.

29. *Green River Star,* 22 May, 12 June, 14 Aug. 1908.

30. Crampton, C. Gregory, and Dwight L. Smith, eds., *The Hoskaninni Papers, Mining in Glen Canyon 1897–1902,* University of Utah Anthropological Papers No. 54, (Salt Lake City: University of Utah Press, 1961), pp. 71–75.

31. Crampton and Smith, *Hoskaninni Papers,* p. 104; *Arizona Sentinel* (Yuma), 4 June 1898; *United States vs. Utah,* pp. 786, 831, 918.

32. *United States vs. Utah,* p. 851; *Arizona Sentinel* (Yuma), 21 Jan. 1899; Crampton and Smith, *Hoskaninni Papers,* pp. 121–40.

33. Crampton and Smith, *Hoskaninni Papers,* pp. 136–43, 147–49; *United States vs. Utah,* p. 696.

34. *United States vs. Utah,* pp. 828–29, 833, 1249–55, 1259.

35. C. Gregory Crampton, *Standing Up Country* (New York: Knopf and University of Utah Press, 1964), pp. 142–44.

36. *United States vs. Utah,* pp. 657, 686–87, 782–89, 799ff, 810.

37. *Ibid.,* pp. 789–90.

38. *Ibid.,* pp. 688–89, 786–87, 789, 800.

39. *Moab Times-Independent,* 30 Dec. 1971, 24 Feb., 4 May 1972.

Closing the River

1. *Arizona Sentinel* (Yuma), 1 Apr. 1909.

2. *Los Angeles Star,* 9 Apr., 16 July 1859; 17 Mar. 1860; *Alta California* (San Francisco), 9 Apr. 1860; *San Diego Union,* 26 Aug. 1873; *Arizona Sentinel* (Yuma), 25 May 1878, 16 Sept. 1893, 6 Mar. 1897; Otis B. Tout, *The First Thirty Years 1901–1931, Being an Account of the Principal Events in the History of Imperial Valley* (San Diego: Otis B. Tout, 1931), pp. 25–26.

3. *Arizona Sentinel* (Yuma), 14 Aug., 11 Sept. 1897, 2 Apr. 1898.

4. *Ibid.,* 25 Dec. 1897, 15 Jan., 2, 16 Apr., 25 June, 13 Aug. 1898.

The *Gila* above the Southern Pacific Railroad bridge at Yuma in the late 1870s.

5. *Arizona Sentinel* (Yuma), 30 May 1900; Tout, *First Thirty Years* pp. 45–47.

6. Tout, *First Thirty Years,* pp. 33–36; Harry T. Cory, "Irrigation and River Control in the Colorado River Delta," *Transactions of the American Society of Civil Engineers* 76 (1914): 1251–59.

7. Cory, "Irrigation," pp. 1258–63; Tout, *First Thirty Years,* pp. 29–50.

8. Tout, *First Thirty Years,* pp. 29–50.

9. Tout, *First Thirty Years,* p. 36; Cory, "Irrigation," pp. 1291–92.

10. Cory, "Irrigation," pp. 1278–88, 1293; *Arizona Sentinel* (Yuma), 26 Sept. 1891, 27 Aug. 1892.

11. Cory, "Irrigation," pp. 1287–91; Godfrey Sykes, *The Colorado Delta* (New York: American Geographical Society, 1937), p. 115.

12. Tout, *First Thirty Years,* p. 100.

13. Cory, "Irrigation," pp. 1291–92; U.S. Congress, Senate, *Senate Document* 212, "Imperial Valley or Salton Sink Region," 59th Cong., 2d sess. (Washington, D.C.: Government Printing Office 1907), pp. 20–39.

14. *Arizona Sentinel* (Yuma), 11 Oct. 1905; Cory, "Irrigation," pp. 1292–1302, 1328.

15. Cory, "Irrigation," pp. 1292–1302; Tout, *First Thirty Years,* p. 105.

16. *Arizona Sentinel* (Yuma), 11 July, 15, 29 Aug., 7 Nov. 1906; Cory, "Irrigation," pp. 1314–49.

17. Cory, "Irrigation," pp. 1310–12.

18. *Arizona Sentinel* (Yuma), 12 Dec. 1906, 15 Feb. 1907.

19. *Senate Document* 212, 59 Cong., 2 sess., pp. 10–11, 17.

20. Cory, "Irrigation," pp. 1356–67.

21. Cory, "Irrigation," pp. 1404–11, 1431–32; Tout, *First Thirty Years,* p. 109.

22. *Annual Report of the Reclamation Service* (Washington, D.C.: Government Printing Office, 1904), pp. 192–93; and *Annual Report of the Reclamation Service* (Washington, D.C.: Government Printing Office, 1905), pp. 97–99.

23. *Annual Report* (1905) pp. 97–99; Cory, "Irrigation," pp. 1235–43.

24. Cory, "Irrigation," pp. 1235–43; *Annual Report of the Reclamation Service* (Washington, D.C.: Government Printing Office, 1910), pp. 72–82.

25. *Arizona Sentinel* (Yuma), 1, 15 May 1907, 25 Mar. 1909.

26. *Ibid.,* 11, 25 Mar. 1909.

27. *Yuma Sun,* 17 June 1910; *Annual Report of the Reclamation Service* (Washington, D.C.: Government Printing Office, 1916), p. 74; "Yuma Project Annual History 1916," p. 15, manuscript in Record Group 115, "Records of the Bureau of Reclamation," U.S. National Archives and Records Service, Washington, D.C.

28. *San Diego Union,* 20 Dec. 1924; *Arizona Republican* (Phoenix), 17 Jan. 1922.

The *Searchlight* at the Imperial canal intake below Yuma about 1905.

Bibliography

Unpublished Material

Bancroft Scraps, "Arizona Miscellany" and "Southern California, San Diego." Bancroft Library, University of California, Berkeley.

Coolidge, Richard Norman. "History of the Colorado River During the Steamboat Era." M.A. thesis, San Diego State College, 1963.

Cory Family Papers. Special Collections Library, University of California, Los Angeles.

Dolley Collection. H. E. Huntington Library and Art Gallery, San Marino, California.

Fleming, Frances. "The History of Steam Transportation on the Colorado River." M.A. thesis, Arizona State College at Tempe, 1950.

Force, Edward Truesdell. "The Use of the Colorado River in the United States, 1850–1933." Ph.D. thesis, Univ. of Calif., Berkeley, 1936.

Forest, Mary Rose. "Yuma the Gateway to California, 1846–1877." M.A. thesis, University of California, Berkeley, 1946.

Hayden Collection. Arizona Pioneers' Historical Society, Tucson.

Hayes, Benjamin, "Emigrant Notes," Part 4 and Scrapbooks "Colorado River" and "Navigation of the Colorado." Bancroft Library, University of California, Berkeley.

Holliday Collection. Arizona Pioneers' Historical Society, Tucson.

Johnson, George A. "Life of Captain George A. Johnson," typescript. California State Library, Sacramento.

———. Statement, 1881, Bancroft Library, Univ. of Calif., Berkeley.

Johnson Papers, Correspondence of Charles, George and Albert Johnson 1813–1900, Arizona Pioneers' Historical Society, Tucson.

Marston Collection. H. E. Huntington Library and Art Gallery, San Marino, California.

Robertson, Frank D. "A History of Yuma, Arizona, 1540–1920." M.A. thesis, University of Arizona, 1942.

Stanton, Robert Brewster, "The River and the Canyon." Manuscript Division, New York Public Library.

United States v. Utah, Complainant's Exhibit 624, Church Historian's Office, Salt Lake City.

U.S., Bureau of Reclamation Records Group 115, U.S. National Archives and Records Service, Washington, D.C.

Published Material

Alexander, J. A. *The Life of George Chaffey.* Melbourne and London: Macmillan & Co., 1928.

Allen, D. K. "The Colorado River." *Arizona Magazine* 2 (Aug. 1893):59–68.

"The Arizona Fleet." *Arizona Sentinel,* 28 Sept. 1878.

Arrington, Leonard J. "Inland to Zion: Mormon Trade on the Colorado River, 1864–67." *Arizona and the West* 8 (1966):239–50.

Ashbaugh, Don. *Nevada's Turbulent Yesterdays*. Los Angeles: Westernlore Press, 1963.
Avillo, Philip J., Jr. "Fort Mohave: Outpost on the Upper Colorado." *Journal of Arizona History* 11 (1970):77–88.
Bandel, Eugene. *Frontier Life in the Army, 1854–1861*. Glendale, Calif.: Arthur H. Clark Co., 1932.
Barney, James M. "La Paz, A Famous Camp on the Colorado River." *Arizona Highways* 15 (July 1939):14–15, 40.
———. "Steamboats on the River." *Arizona Highways* 28 (Feb. 1952):2–5.
Battye, Charles. "Colorado River Days," *Arizona Highways* 12 (Dec. 1936):7, 22–23.
———. "Puerto Ysabel, Ghost Port of the Gulf." *Arizona Highways* 11 (Sept. 1935):8–9, 20–22.
Bear River City, Wyo., *Frontier Index*, 1868.
Beattie, G. W. "Diary of a Ferryman and Trader at Fort Yuma, 1855–1857." *Annual Publications of the Historical Society of Southern California* 14 (1928/29):89–128, 213–42.
Berton, Francis. *A Voyage on the Colorado – 1878*. Ed. and trans. by Charles N. Rudkin. Los Angeles: Glen Dawson, 1953.
Biddle, Ellen McGowan. *Reminiscences of a Soldier's Wife*. Philadelphia: J. B. Lippincott, 1907.
Blanchard, C. J., "Fighting for the Mastery of a Great River," *Ridgway's*, 26 Jan. 1907, pp. 6–7.
Browne, J. Ross. *Adventures in Apache Country: A tour through Arizona and Sonora, with Notes on the Silver Regions of Nevada*. New York: Harper & Bros., 1869.
Bufkin, Donald. "The Lost County of Pah-Ute." *Arizoniana* 5 (Summer 1964):1–11.
Carlson, Edward. "Martial Experiences of the California Volunteers." *Overland Monthly* 7 (May 1886):480–96.
Casebier, Dennis G. *Camp El Dorado, Arizona Territory, Soldiers, Steamboats and Miners on the Upper Colorado River*. Tempe, Arizona: Arizona Historical Foundation, 1970.
Central Pacific Railroad and Leased Lines, Official List of Officers, Stations Agents, Table of Distances, etc. San Francisco: Auditor's Office, 1 March 1879.
Cheyenne, Wyo., *Sun Annual*, 1 Nov. 1892.
Chicago, *Western Mining World*, July 1902–June 1903.
Childers, Martin. "The Ghosts of Port Isabel." *Desert Magazine* 29 (Oct.–Nov. 1966):25–26.
Colton, Harold S. "The Colorado Steamboat 'Charles H. Spencer,'" *Steamboat Bill*, March 1957, pp. 6–7.
———. "Steamboating in Glen Canyon of the Colorado River," *Plateau* 35 (1962):57–59.
Cory, Harry T. *The Imperial Valley and Salton Sink*. San Francisco: J. J. Newbegin, 1915.
———. "Irrigation and River Control in the Colorado River Delta." *Transactions of the American Society of Civil Engineers* 76 (1914):1204–1571.
Cowan, Robert Ernest. "Bancroft's Guide to the Colorado Mines," *California Historical Society Quarterly* 12 (1933):3–10.
Crampton, C. Gregory. *Standing Up Country, The Canyon Lands of Utah and Arizona*. New York: Alfred A. Knopf and Salt Lake City: University of Utah Press, 1964.
Crampton, C. Gregory and Smith, Dwight L., eds. *The Hoskaninni Papers, Mining in Glen Canyon 1897–1902*. University of Utah Anthropological Papers No. 54. Salt Lake City: University of Utah Press, 1961.
Crofutt's New Overland Tourist and Pacific Coast Guide. Chicago: Overland Pub. Co., 1878–79.
Dellenbaugh, Frederick S. *The Romance of the Colorado River*. New York: G. P. Putnam's Sons, 1909.
Denver, *Colorado Sun*, 3 July 1892.
Denver, *Republican*, 4 June 1893.
Dolley, Frank S. "Wife at Port Isabel," *Westerners' Brandbook (Los Angeles Corral) 1957*, pp. 271–85.
Drago, Harry Sinclair. *The Steamboaters*. New York: Bramhall House, 1967.
Dunham, Dick, and Dunham, Vivian. *Our Strip of Land, A History of Daggett County, Utah*. Manila, Utah: Daggett County Lions Club, 1947.
Duryea, Edwin, Jr. "The Salton Sea Menace." *Out West* 24 (1906):3–24.
Edwards, James L. "Ehrenberg, Ghost Town on the Colorado." *Arizona Highways* 9 (Oct. 1933):6–7, 21–22.
Ekker, Barbara Baldwin. "Freighting on the Colorado River," *Utah Historical Quarterly* 32 (1964):122–29.
Farquhar, Francis P. *The Books of the Colorado River and the Grand Canyon*. Los Angeles: Glen Dawson, 1953.
Faulk, Odie B. "The Steamboat War that Opened Arizona." *Arizoniana* 5 (Winter 1964):1–9.

Gilmore, G. W. "Report of a Trip up the Colorado in November 1866 with Captain Rogers on the Esmeralda." In J. Ross Browne, *Resources of the Pacific Slope*. Washington, D.C.: Government Printing Office, 1867.

Gorby, J. S. "Steamboating on the Colorado." *Touring Topics* 20 (July 1928): 14–18, 45.

Green River, Wyo., *Star*, 1908.

Guild, William G., *Arizona*, Washington, D.C.: GPO, 1891.

Guinn, J. M. "Glanton War." *Annual Publication of the Historical Society of Southern California* 6 (1903): 50–62.

Hall, Sharlot M. "The Problem of the Colorado River," *Out West*, 25 (1906): 305–32.

Haskett, Bert. "Steam Navigation on the Colorado River," *Arizona Highways* 11 (Mar. 1935): 6–7, 17–18.

Hinton, Richard J., *The Handbook to Arizona*. San Francisco: Payot, Upham and Co., 1878.

History of Arizona Territory. San Francisco: Elliott and Co., 1884.

Hodge, Hiram C. *Arizona As It Is, Or the Coming Country*. New York: Hurd and Houghton, 1877.

Hosmer, Helen. "Imperial Valley," *American West* 3 (1966): 34–49, 79.

Howe, Edgar F., and Hall, Wilbur Jay. *The Story of the First Decade In Imperial Valley*. Imperial, Calif.: Howe and Sons, 1910.

Hundley, Norris C. *Dividing the Waters: A Century of Controversy Between the United States and Mexico*. Berkeley and Los Angeles: University of California Press, 1966.

Hunter, Milton Reed. "The Mormons and the Colorado River." *American Historical Review* 44 (1939): 549–55.

Ingersoll's Century Annal of San Bernardino County 1769 to 1904. Los Angeles: Ingersoll, 1904.

Jeffrey, John Mason. *Adobe and Iron: The Story of the Arizona Territorial Prison at Yuma*. La Jolla, Calif.: Prospect Avenue Press, 1969.

Johnson, Charles Granville. *History of the Territory of Arizona and the Great Colorado of the Pacific*. San Francisco: Vincent Ryan and Co., 1868.

Johnson, George A. "Exploration of the Colorado River," *Golden Era* 37 (1888): 216–18.

———. "The Steamer General Jesup," *Quarterly of the Society of California Pioneers* 9 (1932): 108–13.

Kemble, John Haskell. "To Arizona by Sea, 1850–1877." *Westerners' Brandbook* (Los Angeles: L.A. Corral, 1963), pp. 137–52.

Kennan, George. *The Salton Sea: An Account of Harriman's Fight with the Colorado River*. New York: Macmillan, 1917.

Leavitt, Francis Hale. "Steam Navigation on the Colorado River." *California Historical Society Quarterly* 22 (1943): 1–25, 151–74.

Lingenfelter, Richard E. *First Through the Grand Canyon*. Los Angeles: Glen Dawson, 1958.

Lippincott, J. B. "The Yuma Project," *Out West* 20 (1904): 505–18.

Lockwood, Frank C. "Steamboat Captain on the River." *Desert Magazine* 4 (June 1941): 13–16.

Long, Paul. "Mineral Park: Mohave County," *Arizoniana* 3 (Summer 1962): 1–8.

Los Angeles, *Mining Review*, 1900–1903.

Los Angeles, *Southern News*, 1863–1864.

Los Angeles, *Star*, 1851–1863.

Love, Frank. *Mining Camps and Ghost Towns: A History of Mining In Arizona and California along the Lower Colorado*. Los Angeles: Westernlore Press, 1974.

MacDougal, D. T. "Voyage Below Sea Level on the Salton Sea." *Outing* 51 (1908): 592–601.

MacHunter, Audrey and Randall Henderson. "Boom Days in Old La Paz," *Desert Magazine* 21 (Sept. 1958): 19–21.

MacMullen, Jerry. *Paddle-Wheel Days in California*. Stanford, Calif.: Stanford University Press, 1944.

———. "Self Liquidating Shipyard," *Westways* 33 (Sept. 1941): 24–25.

Marston, Otis. "River Runners: Fast Water Navigation." *Utah Historical Quarterly* 28 (1960): 291–308.

Martin, Douglas D. *Yuma Crossing*. Albuquerque: University of New Mexico Press, 1954.

Merchant Vessels of the United States. Washington, D.C.: GPO, 1895–1918.

Miller, David H. "The Ives Expedition Revisited, A Prussian's View," *Journal of Arizona History* 13 (1972): 1–25.

Mills, Hazel Emery. "The Arizona Fleet." *American Neptune* 1 (1941): 255–74.

Moab, Utah, *Grand Valley Times*, 1896–1911.

Moab, Utah, *Times-Independent*, 1971–1972.

Mowry, Sylvester. *Arizona and Sonora*. New York: Harper Bros., 1864.

Myrick, David F. *Railroads of Arizona*. Berkeley: Howell-North, 1975.

———. *Railroads of Nevada and Eastern California*. Berkeley: Howell-North, 1963.

Nadeau, Remi A. *The Water Seekers*. Garden City: Doubleday, 1950.
Needles, Calif., *Needles Eye*, 1889–1894, 1906–1914.
Nevada, *Biennial Report of the State Mineralogist of the State of Nevada for the Years 1871 and 1872*. Carson City, 1872.
New York, *Engineering and Mining Journal*, 1901–1902.
Pacific and Colorado Steam Navigation Company. *The Colorado River and Its Relation to the Commerce of San Francisco*. San Francisco, 1865.
Payson, A. H. "Examination and Survey of the Colorado from Fort Yuma to Eldorado Canyon." *Annual Report of Engineers, 1879*. Washington, D.C.: Government Printing Office, 1879.
Phoenix, *Arizona Gazette*, 27 June 1893, 10 Apr. 1895.
Phoenix, *Arizona Republican*, 9 Apr. 1895, 1 Jan. 1896, 17 Jan. 1922.
The Piute Company of California and Nevada. San Francisco: E. Bosqui and Co., 1870.
Prescott, *Arizona Miner*, 1864, 1867, 1873–1874.
Raymond, Rossiter W., *Mineral Statistics West of the Rocky Mountains*. Washington, D.C.: Government Printing Office, 1870.
Riggs, John L. "William H. Hardy: Merchant of the Upper Colorado." *Journal of Arizona History* 6 (1965):177–87.
Robinson, R. E. L. "Yuma, Arizona." *California Illustrated Magazine* 4 (1893):868–78.
Rockwood, Charles Robinson. *Born of the Desert*. Calexico, Calif.: Calexico Chronicle, 1930.
Rusho, Wilbur L. "Charlie Spencer and His Wonderful Steamboat." *Arizona Highways* 38 (Aug. 1962):34–39.
Sacramento, Calif., *Daily Union*, 30 Aug. 1855, 29 Sept. 1862.
Salt Lake City, *Deseret News*, 29 Mar. 1865, 11 July 1866, 16 June 1869.
Salt Lake City, *Daily Telegraph*, 3 Jan. 1865.
San Diego, Calif., *Herald*, 1852–1855.
San Diego, Calif., *Daily Union*, 9 July 1862, 18 July 1885.
San Francisco, *Alta California*, 1852–59, 1864–66, 1870, 1872.
San Francisco, *Evening Bulletin*, 1 Oct. 1867.
San Francisco, *California Express*, 23 July 1864.
San Francisco, *Democratic Press*, 1 Aug. 1864.
San Francisco, *Herald*, 1851–1852, 1858.
San Francisco, *Mining and Scientific Press*, 1864–1910.
Santa Fe, *Gazette*, 10 May 1856.
Schonfeld, Robert G. "The Early Development of California's Imperial Valley," *Southern California Quarterly* 50 (1968):279–307, 395–426.
Searchlight, Nev., *Bulletin*, 1906–1910.
Searchlight, Nev., *Searchlight*, 1905–1906.
Smith, Melvin T. "Colorado River Exploration and the Mormon War." *Utah Historical Quarterly* 38 (1970):207–23.
Stegner, Wallace. *Beyond the Hundredth Meridian*. Boston: Houghton Mifflin Co., 1954.
Summerhayes, Martha. *Vanished Arizona*. Philadelphia: Lippincott, 1908.
Sweeny, Thomas William. *Journal of Lt. Thomas W. Sweeny, 1849–1853*. Arthur Woodward ed. Los Angeles: Westernlore Press, 1956.
———. "Military Occupation of California 1849–1853." *Journal of the Military Service Institution of the United States* 46 (1909):97–117, 267–89.
Sykes, Godfrey. *The Colorado Delta*. New York: American Geographical Society, 1937.
Tout, Otis B. *The First Thirty Years 1901–1931, Being an Account of the Principal Events in the History of Imperial Valley*. San Diego: Otis B. Tout, 1931.
Townley, John M. "Early Development of Eldorado Canyon and Searchlight Mining Districts," *Nevada Historical Society Quarterly* 11 (Spring 1968):5–25.
"Trip to the Colorado Mines in 1862," *California Historical Society Quarterly* 12 (1933):11–17.
Tucson, *Arizona Star*, 23 Dec. 1908.
Tucson, *Arizonian*, 14 Mar. 1869.
Turnbull, Belle. "Gold Boats on the Swan." *Colorado Magazine*, 39 (1962):241–62.
Tuttle, Edward D. "The Colorado River." *Arizona Historical Review* 1 (1928):50–68.
United States v. Utah; Abstract in Narrative Form of the Testimony Taken before the Special Master, 3 vols. Washington, D.C.: GPO, 1931.
U.S. Congress. House. *House Document* 246. 71st Cong. 2d sess. "Judgements Rendered by the Court of Claims." Washington, D.C.: GPO, 1931.
U.S. Congress. House. *House Document* 273. 74th Cong., 2d sess. "Walapai Papers." Washington, D.C.: GPO, 1936.
U.S. Congress. House. *House Document* 953. 61st Cong., 2d sess. "Grand and Green Rivers, Utah." Washington, D.C.: GPO, 1911.

U.S. Congress. House. *House Executive Document* 1, pt. 2. 46th Cong., 2d sess. "Report of Chief of Engineers, Appendix JJ." Washington, D.C.: GPO, 1879.

U.S. Congress. House. *House Executive Document* 18. 51st Cong., 2d sess. "Preliminary Examination of the Colorado River, Arizona." Washington, D.C.: GPO, 1890.

U.S. Congress. House. *House Executive Document* 67. 56th Cong., 2d sess. "Preliminary Examination of the Colorado River, Nevada." Washington, D.C.: GPO, 1900.

U.S. Congress. House. *House Executive Document* 81. 32nd Cong., 1st sess. "Report of a Reconnaissance of the Gulf of California and the Colorado River." Prepared by George H. Derby. Washington, D.C.: GPO, 1852.

U.S. Congress. House. *House Executive Document* 90. 36th Cong., 1st sess. "Report upon the Colorado River of the West, Explored in 1857 and 1858." Prepared by Joseph C. Ives. Washington, D.C.: GPO, 1861.

U.S. Congress. House. *House Executive Document* 124. 35th Cong., 1st sess. "Report of Edward Fitzgerald Beale to the Secretary of War Concerning the Wagon Road from Fort Defiance to the Colorado River." Washington, D.C.: GPO, 1855.

U.S. Congress. House. *House Executive Document* 135. 34th Cong., 1st sess. "Report of the United States and Mexican Boundary Survey." Prepared by William H. Emory. Washington, D.C.: GPO, 1857.

U.S. Congress. House. *House Executive Document* 166. 42d Cong., 2d sess. "Freight to Salt Lake City by the Colorado River." Washington, D.C.: GPO, 1872.

U.S. Congress. House. *House Miscellaneous Document* 12. 41st Cong., 3d sess. "The Exploration of the Colorado River and Its Tributaries." Prepared by Samuel Adams. Washington, D.C.: GPO, 1871.

U.S. Congress. House. *House Miscellaneous Document* 204, 58th Cong., 2d sess. "Preliminary Examination of the Colorado River from Yuma to the Mexican Boundary Line." Washington, D.C.: GPO, 1903.

U.S. Congress. House. *House Report* 1936. 61st Cong., 3d sess. "Southern Pacific Imperial Valley Claim." Washington, D.C.: GPO, 1911.

U.S. Congress. Senate. *Senate Document* 13. 89th Cong., 1st sess. "Federal Census — Territory of New Mexico and Territory of Arizona." Washington, D.C.: GPO, 1965.

U.S. Congress. Senate. *Senate Document* 212. 59th Cong., 2d sess. "Imperial Valley or Salton Sink Region." Washington, D.C.: GPO, 1907.

U.S. Congress. Senate. *Senate Executive Document* 2, pt. 2. 36th Cong., 1st sess. "Report of the Secretary of War, Affairs in Department of California." Washington, D.C.: GPO, 1861.

U.S. Congress. Senate. *Senate Executive Document* 7. 30th Cong., 1st sess. "Notes of a Military Reconnaissance from Fort Leavenworth in Missouri to San Diego, California . . . made in 1846–7." Prepared by William H. Emory. Washington, D.C.: GPO, 1848.

U.S. Congress. Senate. *Senate Executive Document* 51. pt. 10. 50th Cong., 1st sess. "Testimony, Reports of Accountants and Engineers and of the Commissioners of the United States Pacific Railway Commission." 8 vols. Washington, D.C.: GPO, 1888.

U.S. Congress. Senate. *Senate Executive Document* 59. 33d Cong., 1st sess. "Report of an Expedition Down the Zuni and Colorado Rivers." Prepared by L. Sitgreaves. Washington, D.C.: GPO, 1854.

U.S. Reclamation Service. *Annual Report.* Washington, D.C.: GPO, 1902–16.

War of the Rebellion; a Compilation of the Official Records of the Union and Confederate Armies. Ser. 1, vol. 50, pt. 1. Washington, D.C.: GPO, 1897.

Weight, Harold O. "California's First Gold — at the Pot Holes." *Westerners' Brandbook* (Los Angeles: L.A. Corral, 1950), pp. 17–22.

Wells, A. J. "Capturing the Colorado." *Sunset* 18 (1907): 391–404.

Wheeler, George M. *Annual Report upon the Geographical Surveys.* Appendix JJ. Washington, D.C.: GPO, 1876.

White, Lovell. "El Rio Colorado del Sur." *Overland Monthly* 9 (1872):360–70.

Winther, Oscar Osburn. *The Transportation Frontier, Trans-Mississippi West 1865–1890.* New York: Holt, Rinehart and Winston, 1964.

Woodbury, David Oakes. *The Colorado Conquest.* New York: Dodd, 1941.

Woodward, Arthur. *Feud on the Colorado.* Los Angeles: Westernlore Press, 1955.

Yates, Richard. "Locating the Colorado River Mission San Pedro y San Pablo de Bicuner." *Journal of Arizona History* 13 (1972):123–30.

Yuma, *Arizona Sentinel,* 1872–1911.

Yuma, *Morning Sun,* 17 July 1910.

Caged in against mosquitoes, the *Searchlight,* operated by the U.S. Reclamation Service until 1916, became the last survivor of the Colorado River steamboat fleet.

Index

(Illustrations indicated by *italics* and maps by an asterisk)

Acapulco, 91
Aciquia, 167
Adams, Samuel ("Steamboat"), 43–51, 105, 159
Advance, 80*, 100–101, 127, 156–*157*
Advance Gold Dredging Co., 99
Alamo Ranch, 168
Alamo River, 10*, 135, 138*–39, *141*–42
Algodones, 34*, 60, 167
Alpha, 137, 139, 142, *144*
Altar, 76
Alturas, 35
Alvarado, Estefana, 40
Alvarez Ranch, 168
American Camp, 36
Anderson, Charles, 115
Anderson, John A., 97
Andrade, Guillermo, 60, 74, 76, 139
Ankrim, William J., 5

Arctic Landing, 168
Arizona, 47
Arizona and California Railroad, 80*, 97
Arizona City (Yuma), 10*, 15, 34*, 41, 60, 80*, 167
Arizona Navigation Co., 49, 164
Arizona Territorial Prison, 62, 136
Armistead, Lewis A., 24
Arno, 33
Atlantic and Pacific Railroad, 80*, 82–83
Aubrey, Francois Xavier, 66
Aubrey City, 34*, 40, 62, 66, 68, 81, 168
Auger Riffle, 106
Aztec, 86–91, *87, 89*
"Aztec Territory," 68

Babson, John W., 91–*92*
Baby Black Eagle, 118

Bahia San Luis Gonzaga, 88
Baja California, 12, 86, 88
Baldwin, Clarence, 118
Barge No. 1, 44, 51; *No. 2,* 44, 51, *63,* 73; *No. 3, 30,* 44, 51, 82; *No. 4,* 44–45
Barges, 23, *30,* 35, 43–45, 47, 51, *68,* 76, 86, 91, 97, 124, 129, 145, *160*
Beal, 169
Beale, Edward F., 19, 23
Beale's Crossing, 10*, 21, 23–24, 169
Beaver Island, 168
Bennett, Frank, 127
Bermingham, John, 74, 76
Bessie May, 97
Beta, 142, *143,* 144
Big Boat, 118–*119*
Big Sandy Creek, 34*, 68
Bill Williams' Fork, 10*, 44, 66, 71, 97
Black, Robert, 84

188 *Index*

Black Canyon, 10*, 21, 33, 49, 78
Black Crook, 44, 51
Black Eagle, 117, 161
Black Metal Ledge, 84
Black Point, 168
Blaisdell, E. S., 88
Blake, Henry E., 118
Blythe, Thomas H., 60, 74, 76, 81
Blythe Landing, 80*, 81, 84, 86, 168
Boat Knee Bend, 167
Boat Rock, 168
Boulder Canyon, 78, 80*
Bowman, Mrs., 15
Bradshaw's Ferry, 37, 168
Breckenridge, 99
Brown, G. W., 168
Browne, J. Ross, 31
Buckskin Hills, 66
Bucyrus Company, 124
Buena Vista, 167
Bullion Bar Dredging Co., 100
Busby, Erven E., 91

C. M. Weber, 76
Cairook, 24, *27*
Caldwell, H. J., 99
Calexico, 138*, 145
California and Mexican Steamship Line, 74
California Camp, 65, 168
California Development Co., 139–45, 149, 159, 165
California King Gold Mines Co., 84, 95
California Steam Navigation Co., 40

Call, Anson, 47–*48*
Calloway, O. P., 81
Callville, 34*, 47, 49–*50*, 68, 78, 169
Camp Alexander, 169
Camp Colorado, 34*, 66, 168
Camp Independence, 2
Camp Leon, 168
Camp Stone, 108*, 124–*25*
Campo de Ferrar, 36
Campo en Medio, 36
Canyon King, 129, *133*
Capacity, 9, 11
Cargo Muchacho mine, 81, 84
Cargo Muchacho Mining District, 81
Carissa Arroya, 168
Carroll's Creek, 168
Castle Dome Landing, 34*, 40, 62, *67,* 80*, 84, 86, 136, 167
Castle Dome Mining District, 34*, 40, 62, 80*–81
Castle Dome Mountains, 40
Castle Dome Smelting Co., 62
Cerbat, 34*, 68
Cerbat Range, 39, 68
Chaffey, George, 136, 139, 142
Chandler, Richard D., 51
Charles H. Spencer, 129, *130–32*, 161
Chemehuevi Indians, 23, 66, 81
Chemehuevis Valley, 68, 80*, 84
Chimawavo Mining District, 40
Chimehuevis Landing, 34*, 68, 168
Chimney Peak, 10*, *18*
Chimney Peak Landing, 64, 167
Chims Valley, 168

Chinese, 62, 74, 82
Chloride, 34*, 68
Chocolate Mountains, 40, 62
Cisco, 108*, 112
City of Moab, 113–15, *114*
Cliff Dweller, 115–17, *116,* 161
Clip, 80*, 84, 168
Clip Mining Co., 81
Clover, H. K., 113
Cochan, 53, 91, *93,* 95, 97, 99, *102, 134,* 142, 156, 159, 161
Cocopah (I), 29, 33, 41–*42*, 44, 48, 53, 161
Cocopah (II), *30,* 51–*52,* 66, 73, 76, 161
Cocopah Indians, 1, 5, 9, 12, 17, 51, 60
Cocopah Mountains, 34*, 60
Colonia Lerdo, 34*, 60, 74, 76, 80*, 86
Colorado (I), 16, 24, 29, 41, 162
Colorado (II), 41, *46,* 51, *54,* 73, 76, 162
Colorado (III), 117–18
Colorado (barge), 35
Colorado City, 10*, 15, 34*–35, 167
Colorado Fuel and Iron Co., 117
Colorado Hotel, 12
Colorado River: bridges, *72*–73, 82–*83, 87, 152, 160, 178;* difficulties of navigation, 9, 12, 16, 19, 64–65, 78, 88, 97, 112–13, 115, 129, *153;* estuary, 1, 5, 7*–9, 12, *14,* 16, 33, 41, 43–44, 51,

60–61*, 86, 167; exploration, 16–21; floods, 36, 82, 106, 136, 138*, 142–50, 146–47, 154–55; head of navigation, 16, 19, 21, 39, 53, 68, 76; irrigation, 135–59; maps, 7*, 10*, 34*, 61*, 80*, 108*, 138*; tidal bore, 5, 44, 51
Colorado River and Gulf Transportation Co., 88, 165
Colorado River Dredging Co., 100
Colorado River Indian Reservation, 34*, 66, 168
Colorado River Transportation Co., 95, 165
Colorado Steam Navigation Co., 51, 53, 57, 66, 71, 73–76, 82, 84, 97, 156, 165
Columbia River, 48, 95
Comet, 118–22, *121,* 162
Coonradt, A. R., 97
Coquille, 76
Corbin, J. N., 110, 112–13, 115
Cory, Harry T., 145–49
Cottonwood Island, 34*–35, 68, 169
Cottonwood Valley, 10*, 19
Crawfords, 168
Crescent City, 108*, 122
Cuchan Gold Mining, Milling and Dredging Co., 99

Davis, Jefferson, 16
Delta, 149, *151–52*
Dent's Landing, 168
Derby, George Horatio, 5, 8, 16

Deseret Mercantile Association, 47
Desert Land Act, 136
Douglass, George M., 57
Dredges, 136–144, 149, *151–52, 156–57;* gold, 99–100, 122, *124–27,* 156
Drift Desert, 34*, 65, 168
Driscoll, Captain, 9, 11
Duff's Ferry, 168

Eastman, C. A., 76
Edinger, F. S., 145
Edwards, William Hiram, 109–10
Ehrenberg, 34*, 37–*38,* 62, 65–66, 71, 76, 80*–81, 86, 95, 97, 168
El Dorado, 35
Eldorado Canyon, 31–35, 34*, 37, 40, 44, 48, 62, 68, *69,* 71, *77*–78, 80*, 82, 84, 88, 99, 100, 169
El Dorado City, 34*–35
Eldredge, Kimball C., 48
Electric, 86
Electric Spark, 86
Elizaldo, Geronimo, 86
El Rio, 80*–81, 138*, 167
Empire Flat, 34*, 66, 80*, 168
Enterprise, 86
Esmeralda, 43, 47–48, 162
Eureka Landing, 34*, 64, 167
Explorer, 16–23, *18, 22,* 33, 48, 162
Explorer Rock, 10*, 21, 169

Fair, James G., 86
Ferrar, Juan, 36

Flatboats, 8–9, 68
Flora Temple Mine, 62
Floyd, John B., 16
Forrester, F. L., 95
Fort Gaston, 10*, 24, 168
Fort Mohave, 10*, 24, *26,* 29, 34*–35, 37, 39–41, 62, 68, 80*, 81–82, 84, 95, 169
Fort Yuma, 2, 5, *6–12,* 10*, 15, 17, 19, 24–*25,* 29, 31, 34*, 37, 41, 73, 80*, 138*, 167
Fraser River, 44
Freeman, Legh R., 68
Freeman Mining District, 34*, 40, 44
Freemansburg, 34*, 68, 169
Freighting costs, 5, 33, 37, 47, 68, 73
Friant, Frank, 95

Gandolfo, John, 97
Garnett, Dick, 24
Gasoline boats, 86–88, *89, 90,* 97–99, *98,* 113–*114,* 117–18, 122, 127, 129
General Jesup, 12–*13,* 16, 19, 21, 23–24, 29, 97, 162
General Patterson, 11
General Rosales, 76, 162
General Viel, 11
General Zaragosa, 76
Gila, 30, 53, *55, 67,* 69, 73, 76, 78, 82–*83,* 86, 91, 135, 145, *160,* 162, *178*
Gila City, 31, 33, 34*, 36

Gila Mining and Transportation Co., 31, 164
Gila River, 2, 4*, 5, 7*, 10*, 11, 31, 86
Glanton, John, 2
Glen Canyon, 108*–9, 122–32
Glen Canyon Dam, 129
Godfrey, Joseph H., 78
Goldroad, 80*, 84
Goldwater Brothers, 37
Gorman, George B., 44
Graham, G. M., 109
Grand Canyon, 23, 84
Grand River, 106, 110. *See also* Colorado River
Grand Turn, 169
Grand Valley Times, 110
Gravel Point, 169
Great Salt Lake, 115
Green Grand and Colorado River Navigation Co., 106, 166
Green-Grand River and Moab Navigation Co., 113, 166
Green River, 105–22, 108*
Green River (Utah), 105–18, *107,* 108*, 124
Green River (Wyo.), 105, 108*, 118–22, *120*
Green River Navigation Co., 118, 166
Greenwood City, 34*, 68
Gridiron Landing, 10*, 12, 29, 34*, 47, 60, 167
Guadalupe, 76
Guaymas, 74, 76

Gulf of California, 5, 76, 84, 88, 91, 95
Gulf of California Steamship Co., 74, 165

Halfway Bend, 169
Hall, C. S., 99, 166
Hall, F. W., 99
Hall's Crossing, 108*, 122, 124
Halverson's Ranch, 108*, 110, 115
Hamblin, Jacob, 21
Hanksville, 124
Hanlon's Ferry, 167
Hanna, Peter, 129
Hardscrabble Canyon, 167
Hardy, R. W. H., 5; map by, 7*
Hardy, William Harrison, 39, 47–48, 78
Hardy's Colorado, 10*, 12, 61, 167
Hardyville, 34*, 39–40, 47–48, 62, 68, 71, 78, 80*, 84, 169
Harriman, E. H., 142, 145, 149
Hartshorne, Benjamin M., 2, 8, 11, 40, 51
Haskell, Thales, 21
Hassayampa Mining District, 40
Hawley, F. L., 95
Hearst, George, 35
Heath, W. A., 106
Heintzelman, Samuel P., 2, 5, 8–9, 16
Heintzelman's Point, 61*, 167
Henry's Fork, 118
Hercules, 98–99
Hind, Thomas J., 145
Hinton Island, 167
Hite, Cass, 122

Hite's Ferry, 108*, 122, 127
Hoffman, William, 23–24
Hogan, H. J., 106
Hood's Landing, 169
Hooper, George F., 15
Hope Landing, 169
Hoskaninni, 108*, 124–27, *125–26*
Hoskaninni Company, 122–27
Howard, H. F., 109
Hualapai Indians, 23, 66, 68
Hualapai Mountains, 39
Hutt, William, 97

Ida B, 118
Idaho, 57
Imperial, 138*, *140*
Imperial Canal, 138*–39, 141–49, *143–44,* 180; break, 138*, 142–49, *150, 154*
Imperial Land Co., 139, 142
Imperial Valley, 138*, 149, 155
Invincible, 5, 12
Iola, 98–99
Iratata Flat, 168
Iretaba, 24, *26,* 66
Iretaba City, 34*, 39, 169
Iretaba Mining District, 39
Isabel, 51
Isabel Slough, 60
Ivanpah, 34*, 71
Ives, Eugene S., 135–36
Ives, Joseph Christmas, 16, *17–22,* 33, 135, 164
Ives' dredge, 136–*137,* 139

Jackson, engineer, 12
Jaeger, Louis J. F., 5, 15, 135
Jaeger City, 10*, 15
Jaeger's Ferry, *6,* 15, 73, 167
Jaeger's Slough, 91
Joaquin's Ranch, 169
Johnson, George Alonzo, 2, 5, *8–9,* 11–51, 74, 159
Johnson, George A., and Co., 11-51, 164
Johnson, Lute H., *104,* 109–10
Johnson's Landing, 168
Johnston, Alexander D., 51, 68
Jones, J. P., 67–68
José Ranch, 167

Kae-as-no-com, 19
Kanakas, 51
Kansas City Gold Dredging Co., 99
Katy Lloyd, 97
Kelley's, 168
Kenty, Daniel, 106

Labyrinth Canyon, 106, 108*, 109
Laguna, 34*, 40, 62, 156, 167
Laguna Dam, 135–36, 138*, 156, *158*
Laguna de la Paz, 31, 33–34*, 36–37, 40–41, 47, 60, 168
Lake Powell, 127
Lamar Brothers (Charles P. and William F.), 91, 97, 165
La Paz, 31, 33–34*, 36–37, 40–41, 47, 60, 168
La Paz Mining District, 37

La Plomosa, 36
Larsen, Holger, 118
Larsen, Marius N., 118, 122
La Sociedad de Irrigacion y Terrenos de la Baja California (The Mexican Company), 139, 142, 149
Las Pozas, 36
Layton, V. F., 97
Lee's Ferry, 108*, 122, 127–29, *128*
Leland Mine, 86
Lerdo Landing, 34*, 60, 76, 80*, 86, 167
Leroux, Antoine, 16, 19
Lichfield and Leland's boat, 97
Light House Rock, 168
Lin, Henry, 99–100
Lincoln, Dr., 2
Linwood, 108*, 118, 122
Little Dick, 88
Little Valley, 110
Liverpool Landing, 34*, 68, 168
Lloyd, O. N., 97
Los Chollos, 36
Los Minos Primeras, 36
Louisville, 35
Lumsden, John J., 113–15

Macedonia Mining District, 39
McClatchy, Tex, 129
McCrackin, Jackson, 66
McDonough, A. N., 53, 57
Major Powell, 104, 106, 109–10, 162
Marengo Mining District, 40
Marguerite, 117
Maricopa Wells, 34*, 76

Marysvale, 108*, 129
Mazatlan, 91
Mazatlan, 11, 74, 76
Mellen, 80*, 82, 97, 169
Mellon, John Alexander, 51, 53, 64, *65,* 74, 78, 82, 84, 86, 88, 91, 95, 97, 145, 156, 159
Mendez, Joaquin, 95
Merrell, Homer, 106
Mesa Bend, 168
Metzger, William, 57
Mexicali, 138*, 145, *146*
Mexican Coast Steamship Co., 91
Mexican-Colorado Navigation Co., 91, 95, 97, 156, 165
Mexican Company, The, 139, 142, 149
Meyers, H. J., 100
Mid-West Exploration Co., 118
Miller's Landing, 168
Mineral City, 34*, 37, 168
Mineral Park, 34*, 68, 78, 80*, 82
Mining: coal, 127; copper, 15, 39–40, 44, 66, 117; gold, 15, 31, 36–37, 40, 62, 64, 66, 81, 84, 86, 99–100, 122–29, *123, 128;* lead, 62, 81; manganese, 117; salt, 35, 78; silver, 31, 33, 35, 40, 44, 66, 78, 81, 84; sulphur, 60
Minturn Slough, 29
Mission la Purisima Concepcion, 5
Mission San Pedro y San Pablo de Bicuner, 33
Moab, 106, 108*, 110–13, 117–18, 129
Moab Garage Co., 118

Mohave (I), 44, *46*, 51, 53, *63*, 162
Mohave (II), 53, *56*, 73, 76, *78–79*, 82, 86, 88, 91, 162
Mohave (III), 97
Mohave and Milltown Railway, 86, 99
Mohave Canyon, 10*, *22*, 29
Mohave City, 34*, 40, 169
Mohave Indians, 19–*20*, 23–24, *27*, 51, 66, 81, 95
Mohave Mountains, 39
Mohave Valley, 10*, 19
Möllhausen, Balduin, 17; sketches by, *3*, *18*, *20*, *22*, *27*
Montana, 57–59, *58*
Moquie Mining Co., 127
Morehead, J. C., 2
Mormon Island, 78
Mormons, 16–17, 21, 23, 68, 78, 105
Moss, John, *32*–33, 37, 39, 66, 68
Moss Lode, 34*, 37
Mt. Mejor, 80*, 86
Mulege, 76
Mullins, 129
Murphyville, 80*, 84, 169
Murray's Mine, 168
Myers' Landing, 168

Navajo, 118
Neahr, David, 12, 64
Needles, 80*, 82–88, 95–99, 169
Needles Navigation Co., 99, 166
Needles Reduction Co., 88
Nevada Point, 169
Newbern, 53, 59, 74

New Era Mill, 35
New River, 34*, 60, 80*, 86, 138*
New York Mountains, 84
Nina Tilden, 44, 48–49, 51, 53, 163
North Dakota, 80*, 100
Norton, Edward, 51
Norton, George W., 81
Norton's Landing, 80*–81, 84, 168

Oatman, 80*, 84, 86
Oatman, Olive, 37
Ogden's Landing, 10*, 12, 34*, 60–61, 167
Oleson, E. J., 88
Olive City (Olivia), 34*, 37, 168
Olympia Bar, 108*, 127
Onward, 53
Oppenheimer, Milton, 117
Osborne's Ranch, 168
Overman, Charles C., 43–44, 51, 74

Pacific and Colorado Steam Navigation Co., 48–49, 164
Pacific City, 81, 168
Paddy Ross, 117
Pah-Ute County, 47
Papago Indians, 36
Parker, 80*, 97, 168
Parker's Landing, 34*, 66, 84, 168
Parsons, Levi, 35
Pascual (Kae-as-no-com), 19
Paymaster's Bend, 168
Peas Ranch, 169
Pedrick's, John, 10*, 12, 34*, 60, 61*, 167

Pedrigal, 168
Philadelphia Silver and Copper Mining Co., 44, 164
Philips Point, 61*
Phillips, engineer, 9
Picacho Landing, 34*, 64, 80*, 86, 95, 167
Picacho mine, 64, 80*
Picacho Mining District, 40, 81, 84
Pierson, William H., 51
Pigeon Ranch, 169
Pilot Knob, 138*–39, 149
Piute City, 34*, 68, *70*
Piute Company of California and Nevada, 71
Piute Indians, *32*, 84
Planet, 34*, 40
Planet mine, 40, 66
Planet Wash, 168
Polhamus, Isaac, Jr., *28*–29, 37, 40, 44, 62, 68, 74–*75*, 78, 82, 84, 86, 88, 91, 95, 97, 159
Polhamus Landing, 78, 80*, 169
Polhamus' Ranch, 68, 169
Poole, William, 51, 60
Poole's Landing, 34*, 60
Porfirio Diaz, 91
Port Famine, 10*, 12, 34*, 60, 167
Port Famine Slough, 44
Port Isabel, 34*, 51–*54*, 61*, 74, 167
Port Libertad, 76
Poston, Charles, 15, 66
Potato Point, 168
Pot Holes, 33, 62, 100, 167

Poverty Bar, 169
Powell, 169
Powell, John Wesley, 105, 118
Prescott, 39, 66
Providence City, 34*, 39
Providence Mountains, 39
Pumpkin Seed, 44
Purdy, Warren G., 88
Pyramid Canyon, 10*, 19
Pyramid Mining District, 39

Quartette, 80*, 84, 95, 169
Quartette Mining Co., 84, 95
Queen City mine, 33, 35
Quetovac, 95
Quien Sabe Ranch, 168

Railroads, 86, 91, 105; competition with steamers, 49, 73, 82, 84, 97, 99; narrow-gauge, 86, 95
Rancho de los Peñasquitos, 40
Randolph, Epes, 145
Raousett de Boulbon, Count, 12, 53
Rattlesnake Point, 168
Red Cloud Mining Co., 80*–81
Redondo, José, 36, 62, 64
Redondo's Ranch, 168
Red Rock Gate, *45*, 64, 168
Reliance Landing, 167
Restless, 40
Retta, 95, 97, *163*
Rice, Hill and Co., 88
Rio Grande Western Railroad, 105, 108*
Rioville, 78, 80*, 169

Riverside, 108*, 117
Riverside Mountain, 168
Riverside Ranch, 168
Roaring Rapids, 34*, 47, 49, 169
Robie, F. C., 97
Robinson, David C., 12, *15*–16, 21, 23, 37, *53*–*54*
Robinson's Landing, 10*, *12*–*14*, 17, 61*, 167
Rock Springs Mining District, 39
Rockwood, Charles, 136, 142, 145
Rogers, Robert T., 49
Rood, William B., 64
Rood's Ranch, 34*, 64, 168
Roosevelt, Theodore, 149
Ross, B. S., 106
Ryland, Richard, 40

Sacaton, 168
Sacramento Mining District, 39
Sacramento River, 29, 40, 43–44, 76
St. Vallier, 88, *92*, 95, 97, 99, *102*, 145, *148*, 156, 163, 170
Sakey, 167
Salton Sea, 138*, 145, *147*
Salton Sink, 10*, 135–36, 145
Sam, William, 62
San Blas, 76
San Diego, 5, 8, 40–41
San Felipe, 61*, 76
San Francisco Chamber of Commerce, 43
San Francisco Mining District, 37

Sanguinetti, E. F., 156
San Joaquin River, 76
San Jorge, 95, 163
San Jorge Bay, 91, 95
San Juan, 35
San Pedro, Los Angeles and Salt Lake Railroad, 80*, 99
San Rafael River, 108*, 113, 115
Santa Ana Mining Co., 88, 95, 165
Sarah, 44
Schultz, Robertson and Schultz, 129
Searchlight, 95–*96*, 99, 145, 149, *153*, 159, 163, *180*, *186*
Searchlight, 80*, 84–*85*, 91, 99
Short and Dirty Rapids, 78
Sierra Nevada, 8
Signal, 34*, 68, 78, 80*–81
Silas J. Lewis, 91, *102*, 145
Silver Mining District, 81
Smith, Alphonso B., 91, 95, 97
Smith, Charles M., 95, 97
Smith, Hualapai, 34*, 60, 167
Snively, Jacob, 31, 33, 81
Sonora, 95
Sonorans, 33, 36, 51, 60
Southern Pacific Railroad, 73, 80*–81, 138*, 142, 145, 156; bridge at Yuma, *72*–73, *87*; closing the canal break, 138*, 145–*50*, *154*; operation of steamers, 74–76, 82–83
Southwestern Mining Co., *77*–78
Sou'Wester, 78
Spalding, F. D., 97

Spencer, Charles H., 127–29, 166
Stacy Brothers (E. E. and O. T.), 86–88, 165
Stanton, Robert Brewster, 109, 122–27, *124*
State of Arizona Improvement Co., 136
Steamboat: crews, 51, 74; earnings, 12, 71, 145; engines, 9, 12, 16-17, 41, 43, 95, 106, 110, 129; excursions, 11, *56*, 84, 86, 88–*89*, 95, 105–6, 109, 113-15, 117, 129; explosions, 12, 117; fares, 57, 62, 74; propeller, 95, 106, 117; run aground, 19, 21, 41, 49, 64–65, 109, 112, 115, 118, 122, 129; shipping rates, 12, 33, 37, 40–41, 47, 57, 68, 71, 81–82; side-wheel, 1, 9, 11, 29, 31, 57, 76; stern-wheel, 8, 11, 16–17, 29, 43, 88, 95, 110, 115, 129; sunk 11, 21, 53, 97, 113, 156
Stevenson's Island, 80*, 84, 167
Stewart, W. M., 68
Stillwater Canyon, 109
Stone's Ferry, 169
Stow, Joseph W., 48
Sulphur Bend, 168
Summeril, Frank H., 110–13, 166
Sunbeam, 122
Swallows' Nest Bend, 168
Swamp Land District No. 310 (Blythe), 81
Swan Lagoon, 168
Swan River, 99

Swansea, 40
Sweeney, William, 97
Sykes, Godfrey, 145

Taylor's Camp, 168
Techatticup mine, 33, 35
Teddy R, 122
Tehuantepec, 91
Tercherrum, *32*
Thorne, Stephen, 51, 74
Thornton, Joe, 97
310 Landing, 81, 168
Tickaboo Bar, 108*, 122, 127
Tidal bore, 5, 44, 51
Tilden, Alphonzo F., 44
Treaty of Guadalupe Hildago, 8
Trueworthy, Thomas E., *43*–51
Tungate, D W., 99
Turnbull, James, 9, 11, 164
Twogood, Williams S., 91, 97

U. S. Reclamation Service, 136, 142, 149, 156, 159, 166
Uncle Sam (ocean steamship), 24
Uncle Sam (river steamer), 1, 9, 11–12, 16, 163
Undine, 110–13, *111,* 163
Union Line, 43, 164
Union Pacific Railroad, 105, 108*, *120*
Upper Camp, 35
Urie, W. T., 100
Utah-Nevada Copper Co., 117

Valentine's Bottom, 108*–9, 115
Van Arsdale, W. W., 81
Vanderbilt, 80*, 84
Veagas, 35
Vegas Wash, 10*, 21, 47, 169
Vice, Martin, 44
Victoria, 43
Vineyard, J. R., 35
Violet Louise, 129
Virgin River, 10*, 16, 19, 21, 35, 53, 68, 78, 169
Virginia, 34*, 68
Vista, 115
Vivian mine, 86
Volcano Lake, 60
Vulture mine, 34*, 40

Walker, Joseph, 33
Warm Creek, 108*, 127, 129–*130*
Water Pearl, 97
Wauba Yuma, 68
Wauba Yuma Mining District, 34*, 39
Weaver, Paulino, 19, 33, 36, 40
Weaver Mining District, 37
Welcome Ranch, 167
Western Development Co., 74
Wharton, Joseph, 78
Wheeler, Arthur, 106, 110
Wheeler's Ranch, 106, 108*–10, 117
White, J. G., and Co., 156
White, James, 49
White, James L., 17, 19, 21

White Fawn, 44, 51
White Hills, 84
Wickenburg, 34*, 66
Wickenburg, Henry, 40
Wilburn, L. C., 35, 40, 68
Wilcox, Alfred H., 5, 8, 11–12, 40, 51, 74, 76
Williams Fork Mining District, 40
Williamsport, 34*, 40–41, 64, 167
Willow Camp, 167
Wilmont, 117
Wilson, Sam, 97
Wimmer, Tom G., 117
Wolverton, Edwin T., 117–18
Wozencraft, Oliver M., 135–36
Wright and Lawrence Mining Co., 97

Yaqui Indians, 36
Yavapai Indians, 66
Yokey, Harry T., 115, 117–18, 127, 166
Young, Brigham, 16, 47
Yuma, 51
Yuma, 15, 34*, 60, 62, 73–74, 76, 80*, 84, 86, 91, 95, 97, 99, 136, 138*, 149; steamer landing, *25, 30, 60,63, 102, 178*
Yuma and Port Isabel Railroad, 76
Yuma-Colorado River Gold Dredging Co., 99
Yuma Ferry, 2, 4*–5, *25, 102*
Yuma Indians, *2–3,* 9, 17, 19, 51, 60, 62